# THE
# ROAD
# TO
# GONDWANA

**Bill Morris** is a writer, documentary filmmaker and musician based in Port Chalmers, New Zealand. He has worked extensively as a wildlife filmmaker for NHNZ, the BBC Natural History Unit and others, and is a regular contributor to *New Zealand Geographic* magazine. His passion for science and stories of the natural world informs all of his work. Bill is also a recording folk rock musician and is currently working on his third album, *In the Limestone Country*, an exploration of immigration in New Zealand.

# THE
# ROAD
# TO
# GONDWANA

## In search of the lost supercontinent

## BILL MORRIS

EXISLE
PUBLISHING

First published 2022

Exisle Publishing Pty Ltd
PO Box 864, Chatswood, NSW 2057, Australia
226 High Street, Dunedin, 9016, New Zealand
www.exislepublishing.com

A CiP record for this book is available from the National Library of Australia.

ISBN 978-1-922539-33-5

Designed by Enni Tuomisalo
Typeset in Noto Serif, 10pt
Printed in China

This book uses paper sourced under ISO 14001 guidelines from well-managed forests
and other controlled sources.

10 9 8 7 6 5 4 3 2 1

*For Chuck Landis, a traveller*
*on the road to Gondwana.*

# CONTENTS

# GEOCHRONOLOGICAL SCALE

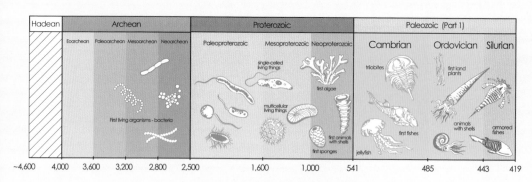

millions of years ago

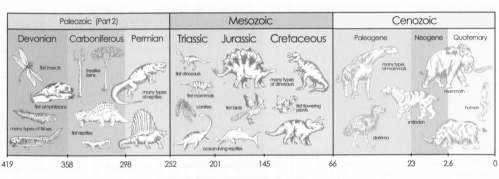

| Paleozoic (Part 2) | | | Mesozoic | | | Cenozoic | | |
|---|---|---|---|---|---|---|---|---|
| Devonian | Carboniferous | Permian | Triassic | Jurassic | Cretaceous | Paleogene | Neogene | Quaternary |

first insects

treelike ferns

many types of reptiles

first dinosaurs

many types of dinosaurs

first mammals

conifers

first birds

first flowering plants

many types of mammals

mammoth

first amphibians

human

many types of fishes

first reptiles

smilodon

diatrima

ocean-living reptiles

419        358        298        252        201        145        66        23        2,6        0

# millions of years ago

# PROLOGUE

In the spring of 1910, the ex-whaling ship *Terra Nova*, every inch of her deck laden with supplies for a polar voyage, sets out on a voyage of exploration that will etch her name, and the names of those on board, into the annals of history.

Crowds in the town of Port Chalmers, in southern New Zealand, line the wharves to watch her go, willing her onwards towards one of the greatest goals of human exploration — the as-yet-unreached South Pole.[1]

Aboard, the expedition's leader, Captain Robert Falcon Scott, contemplates the long journey ahead. As *Terra Nova* rounds Taiaroa Head at the entrance to Otago Harbour, the first Pacific swell tugs at her hull. The men rush to the ropes as cargo shifts in the hold, settling in against the roll and pitch of the ship.

Then the northeasterly grabs her sails and pulls her out onto the sea, away from the green hills and peaceful towns of New Zealand, away from hospitality and comfort, into the howling wilderness.

January 1912, fourteen months later.

A cold sun glares across the Earth's atmosphere, oblivious to the tiny party of emaciated skiers hauling their sledge across the snows of Antarctica — specks on the endless white of the polar ice cap.

These five men — Robert Falcon Scott, Edward Wilson, Lawrence Oates, Edgar Evans and Henry Bowers — are dragging with all their fading strength not only their sledge but also the heavy burden of defeat. Having travelled 1500 kilometres (930 miles) in fierce conditions, they arrived at the South Pole to find a Norwegian flag planted in the featureless ice — their quest to be the first to stand on the Earth's southern axis rendered fruitless by the rival expedition that got there a month earlier.

'Great God this is an awful place,' Scott writes in his diary. 'Now for the run home ... I wonder if we can do it?'[2]

And so now these five men, more alone than any other group of human beings, are pushing their bodies to the limits of exertion to get back to their camp at Cape Evans, on the edge of the Ross Sea; to get away from the ice.

The sastrugi — heaped ridges of wind-blown snow — grip their sledge, sucking the strength from their bones while their food supplies dwindle to dangerously low levels. They are constantly freezing in their heavy, wet gear.

*Scott's polar party near the South Pole, 1912.*

On 4 February, Evans and Scott tumble into a crevasse, with Evans suffering a serious concussion. As the march goes on his condition worsens, and his nose and fingers become severely frostbitten. Oates' feet, too, are frostbitten and Wilson has an injured leg.

The team follow their outbound tracks, aiming in the featureless white for food and fuel depots left on the outward journey, racing against the shortening season. On 8 February the team reaches the top of the Beardmore Glacier — their stairway off the Antarctic Plateau and onto the Ross Ice Shelf, the home stretch. For the first time in weeks they encounter bare rock, something Scott likens to reaching land after a long sea voyage. Despite the pressing urgency of making progress, the explorers decide to stop for an afternoon of 'geologising'.

From the start, this expedition was planned with science at its core — the 'rock foundation of all effort', as recorded in Scott's diary. And so, despite their predicament, Scott and Wilson spend this afternoon exploring ancient sandstone outcrops near Mount Buckley.

The Beacon Sandstone had been discovered on Scott's previous expedition to Antarctica in 1904 — a thick layer of sedimentary rock dissecting the Transantarctic Mountains. Beneath their hammers, the sandstone quickly reveals organic impressions: fossil leaves.[3] This icy wasteland, these fossils say, hasn't always been like this. Antarctica was once warm and fertile, a continent of lush growth and running water. How appealing the idea must seem to the travellers in this frozen moment.

Scott and his men load up the fossils and the next day continue their long march, stopping for half an hour the following morning to collect more fossils from the glacial moraine. They still have the treacherous Beardmore Glacier to negotiate and, after that, 700 kilometres (435 miles) of hard slog across the Ice Shelf.

On the descent down the glacier, a series of wrong turns amid the jumble of broken ice and crevasses quickly changes the game for Scott and his men. If things were tight before, this has now become a race against death. On Monday 12 February, Scott writes: 'In a very critical situation ... Pray God we have fine weather tomorrow.'

Reaching the bottom of the glacier the team begin the long haul across the Shelf. One by one they make the supply depots, finding all of them depleted of food and fuel.

Then, the weather turns hostile. The men find themselves hauling into a constant head wind in temperatures as cold as -40°C (-40°F), putting them at extreme risk of hypothermia. Despite the horrific conditions and with the injured members of their party slowing the group terribly, they still manage to pull up to 17.5 kilometres (11 miles) each day.

Scott's diary becomes laced with anxiety and a growing sense of doom. His entries now discuss little more than weather, ice conditions, miles covered and food — the only things that matter at this point. References to God become more frequent. 'Sunday March 4th — Things looking very black indeed ...'

Oates is struggling along on severely frostbitten feet, while Evans' condition has continued to steadily deteriorate. Gradually, his mind, along with his willpower and resolve, begins to disintegrate. When he finally collapses in the snow and dies in his tent shortly after, Scott records a sense of relief that the man's suffering has at last ended.

But by now the four remaining men are in not much better shape — they are dangerously malnourished and suffering from scurvy, and all are exhausted. Despite the looming sense of futility, they desperately plug on. Incredibly, even now, with the spectre of death upon them, they don't ditch the cache of fossils they retrieved from the Beardmore Glacier.

On 16 March, Oates at last succumbs to the wilderness. As recorded in Scott's diary, he walks out of the tent, uttering the words, 'I am just going outside, and may be some time', before disappearing forever. And so by 29 March there are just three remaining — Scott, Bowers and Wilson, huddled in their sleeping bags in a pitiful tent, just 17.5 kilometres (11 miles) from the well-stocked depot that could save their lives.

A fierce blizzard imprisons them in their tent for four days, making further progress impossible. In the midst of the storm, Scott writes letters to his family and those of Oates, Wilson, Bowers and Evans. And then, in a hand shaking so much it's barely legible, he writes that he can write no more.

'For God's sake, look after our people' is his last sign-off. The diary is complete.

And then the wind comes driving in hard, snow piles up in drifts against the tent, and beyond the constant shifting of the snow, there is no more movement.

With the coming darkness of winter, the whales abandon the ice for warmer latitudes and the continent slips into the long polar night.

When spring arrives, and the light at last returns, a search party sets out from the expedition base at Cape Evans. The team soon find the lost explorers frozen in their tent. With immense difficulty, they pry Scott's diary from his frozen clasp, hearing the ice in his arm break as they do.[4]

Then they cave in the tent, erect a cross made of skis above the site and leave the bodies where they lie. Retrieving the sledge, along with its precious cargo of fossils, they return to Cape Evans to await the *Terra Nova* and the trip home. They have been on the ice now for almost two years. It's time to go home.

High above my house in Port Chalmers, New Zealand, stands a monument. It's built from Port Chalmers breccia, a chunky, blue-grey rock spewed from the hot guts of the Dunedin volcano around 12 million years ago.[5]

The monument is crowned with an anchor, that age-old symbol of safety and hope amid the chaos and storms of life. It looks out across my hometown, with its busy cargo port, its cluster of colourful hillside houses and its numerous churches and pubs. Clouds spill over the edge of the surrounding hills and big cargo ships creep down the channel, tugs at their stern working like bees to guide them in. Beyond the heads, the Pacific Ocean shimmers in a northeasterly breeze.

When the *Terra Nova* returned to New Zealand in the summer of 1913, the world learned of Scott's demise. In response, the city of Dunedin erected this monument, to mark Scott's last port of call. Today, over a hundred years later, it still stands proudly above the town, observing the goings-on of the busy port below. It has watched a century pass — seen the age of sail and steam fade into history, and watched troop ships carrying the town's young, strong men off to two wars. It's observed the old wooden wharves concreted and stacked with shipping containers and witnessed the rise of the cruise ship industry in an era when people take to sea not to seek intrepid adventure but to be pampered, fed and entertained.

At the monument's base a plaque carries a portion of the final note Scott made out to the world: 'Things have come out against us,' he wrote with charismatic understatement. 'We have been willing to give our lives to this enterprise.'

Today, Scott's Antarctic expedition of 1910–13 is often discussed as a heroic failure. But his expedition was about far more than claiming the glory of being the first to step on the pole. It was primarily propelled by a thirst for knowledge in an age when exploration and science went hand in hand. And viewed in this light, the expedition was a success.

Meteorological data painstakingly and meticulously gathered by the crew is still in use to this day, while the fossils collected by Scott and Wilson on the Beardmore Glacier, and which they refused to abandon even in the jaws of death, would, in time, help to reshape the world in the human imagination.

The bodies of those five brave men are still somewhere out there on the barrier. One day they will emerge, as part of a calving iceberg and finally be deposited on the sea floor around Antarctica — free from the ice at last.

Around the base of the Port Chalmers monument, a small gang of roosters scratches for scraps of food in the grass, while throughout the day tourists pull in to read the inscription. The last words on it are from the Bible, a line from Joshua. 'Your children shall ask their fathers in times to come,' it reads, 'saying, "What mean these stones?"'

*Scott (centre) and his team at the South Pole.*

# 1

# WHAT MEAN THESE STONES?

This is the story of a journey. Actually, it's a story of many journeys, of paths woven through the fabric of Earth's history, and of human history, all of which lead to one semi-mythical and yet completely real place: the lost supercontinent Gondwana.

Gondwana is a place that no longer exists and yet which still connects half the world, because the 3 billion of us who live in Africa, South America, India, Australia, Papua New Guinea, New Zealand, and Arabia spend our lives walking around on what's left of it. But more than that, Gondwana has shaped the world we live in. Many of the species we share Earth with evolved there. Had Gondwana never existed, the planet would be a very different place. The trees of our forests would be different. The animals we live among would not be the same. Had Gondwana not existed, maybe we wouldn't either.

The road to Gondwana took western science hundreds of years to travel. And like Scott's epic haul across the ice of Antarctica, it was a journey jagged with many dead ends and wasted miles. When it was finally realized, Gondwana still remained fuzzy, hard to picture. It is still that way. That's partly because Gondwana was a very big place. Anyone who, like me, has travelled in a bus across the outback of Australia or Patagonia for days on end has an appreciation for how big these landmasses are. And yet Australia and South America were just fragments of that much bigger supercontinent.

Answering the question 'What was life like on Gondwana?' is exceedingly difficult. It's a bit like asking what life is like on Planet Earth. Gondwana was a complex mosaic of ecosystems. There were many aspects to it, many biomes and ecoregions, each different from one another. Think of how many different aspects there are to life on just one of our modern continents — Africa, for example, which has rainforests in its west, grasslands rippling across its eastern flanks, snowcapped mountains and one of the world's great deserts in the north. Then you begin to understand the complexity of the task that faces modern paleontologists as they try to piece together what life on the ancient supercontinent was like.

Also, Gondwana existed for an enormous amount of time, perhaps 500 million years or more. In that amount of time, the planet changed beyond recognition. When Gondwana first formed there was no life on land, and animals in the sea were simple, single-celled organisms. When Gondwana finally broke apart, dinosaurs roamed its forested hills and valleys.

So now we see that there is no single 'Gondwana' we can go back in time to visit. There are thousands of different versions of Gondwana. If we want to visit Gondwana, we have to choose one of these versions — one window of time. And to do so, we need a tour guide — something that lived through that particular window of time, and which can therefore tell us something of what life was like then.

Plants, and the fossils they leave, are perhaps our best tour guide of deep time. Plants form the basis of terrestrial ecosystems. They shape the planet and the atmosphere. They are integral to life on Earth, forming the stage upon which animals like us come and go in quick succession. There are plants alive today whose close relatives grew on Gondwana hundreds of millions of years ago.

Plants record, in their trunks and leaves, evidence of changing climate, and the effect that has had on the Earth's biota through time. This becomes increasingly relevant in our modern world as we face potentially catastrophic climate change. Gondwana, too, experienced rapid climate change, and the plants that grew upon it felt the consequences and recorded it in their passing. It is perhaps no surprise, therefore, that a plant would lead scientists to Gondwana and become fundamental in telling the story of the vanished supercontinent. It would also become my tour guide, as I set off on my own journey to discover Gondwana.

I first met my Gondwanan tour guide on a rainy day in my home city of Dunedin when I wandered into the University of Otago's geology museum. The geology department at Otago looks and feels just like you want a geology department building to. Mount the stone steps, concave now from the racket of a century and half of students clattering up them. Pull open the heavy wooden door, and step inside. Inhale the building's smell — the sombre scent of old paper and chalk dust, pipe smoke and rocks.

The corridors twist and turn in a pleasantly dysfunctional way, often terminating abruptly in a windowless office or a cluttered storeroom. Wood-panelled walls are lined with scientific posters. An examination of the oldest basalts of the Dunedin complex. Results of a drilling project along the Alpine Fault. A new climate record from the mud core of a subantarctic lake. There are nooks within crannies and the ghostly voices of long-vanished professors reverberating in the rafters of empty lecture theatres. Shelves creak with the weight of decades upon decades of acquired knowledge, condensed into journals and bound theses. And of course there are rocks and minerals — drawers and drawers of them, the long geological history of this country mined, quarried, dissected, examined, catalogued and stored.

Below the building, a basement is stuffed with the fossilized bones of ancient whales carved from the surrounding hinterland, their huge jaws and skulls lying unlabelled on sagging tables. The shark-like teeth of long-extinct dolphins gather new antiquity in the dusty shelves.

Down the end of one of the corridors you'll find the museum. It's modest by the standards of these things — there's no T-rex teetering beneath artfully angled ceiling lights, no dramatic dioramas of ancient worlds. It's just one quiet room at the end of an old building, visited for the most part only by students of the department. The displays are in wooden cabinets with glass tops, carefully laid out in sequence of geological time, from youngest to oldest, each specimen identified with a typewritten label that quietly yellows and curls in the weak sun filtering through the windows.

Because this is New Zealand, a land that for much of its history has been more or less underwater, most of what is on display is marine: the shells of molluscs and brachiopods, and the casts of echinoderms, bryozoans, corals and other denizens of ancient seabeds. In the first cabinets are specimens from the most recent times — some of these are just a few million years old. They've been carefully removed from the rock that held them by skilled hands with delicate tools, and the quality of preservation is exquisite; they look like they could have been picked off a beach yesterday. In fact, they're far better preserved than your average beach-cast shell as most of them lived and died in deep, quiet water, out of reach of the turbulent waves. Here, they were slowly buried by a rain of sediment and eventually turned by pressure to rock. Almost every feature of their intricate structures, every whorl and spike, corrugation and hinge, is therefore intact.

For the most part, these are fossils of animals that may still be found in our seas today, or whose close relatives, at least, may be. Moving across the aisle, however, brings you to a much older time, 25 million years back, in the Oligocene epoch, when Earth's climate was far warmer than it is today. While the level of preservation in these fossils is still quite high, the species themselves are unfamiliar. There are some, like giant turban shells, that look sort of like animals you'd still find living in the tropics, but for the most part this is most definitely not the New Zealand ocean we know today.

Along the far end of this row of cabinets we go back deeper still, back 66 million years or more to the Triassic, Jurassic and Cretaceous periods, the time of the dinosaurs, flying reptiles and the great mosasaurs and plesiosaurs. These shells really do look old. They have a sort of strange chunkiness, a heaviness, that somehow speaks of antiquity. They're like the geology department building itself — old in a way that could never be authentically recreated. These fossils have also had a harder time; 66 million years is a long time to sit around and, inevitably, time has rounded off their edges, dulling nature's designs so that the fossils start to look a bit like old sucked sweets.

And then you come to the final row of cabinets. Now we're deep in the world's ancient past. Most of these fossils aren't complete —

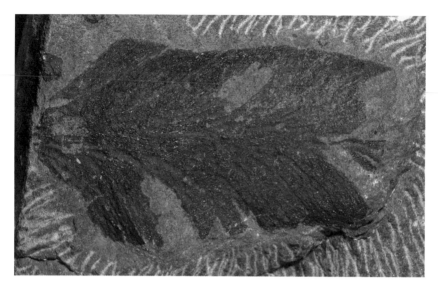

*Glossopteris fossil, University of Otago geology museum, Dunedin, New Zealand.*

some are little more than vague shadows on slaty rock. These come from the Carboniferous, Permian and Devonian periods, the time in Earth's history when the major groups of plants and animals we now recognize were still coalescing from the crucible of evolution.

And there, amid the old shells and fragments of crinoids and graptolites, lies an anomaly. It's one of the most unspectacular fossils in the whole museum — just a crinkled black smudge. You'd likely not even notice it unless you happened to be looking specifically for it. What differentiates it from most of the specimens in this room is that it's not a marine species. It's a land plant. Or, to be exact, the scrappy remains of just one leaf.

The label gives little away, providing only the genus — *Glossopteris spp.* It's a pathetic fragment of a long-vanished forest, washed off the land into shallow marine waters to be preserved in mud on the sea floor. It was chipped and hammered from solid rock, 260 million years later, in a remote gully in southwest New Zealand. Amid the ornate shells and heavy whalebones that fill this room, you'd be forgiven for overlooking this small specimen. But fossils are often about more than just appearances. Sometimes it's the story the fossil has to tell that is far more spectacular. And few have as dramatic a story to tell as this one. *Glossopteris* is a fossil that speaks of no less than the movement of continents across the face of the Earth, and the formation and destruction of the supercontinent Gondwana.

It's a story that has only been pieced together in the last two centuries by men and women, all around the world, who have pored over rock specimens in the dusty bowels of buildings just like this; and who have taken up their hammers and wandered the rocky world, chipping away at the boulder face of understanding until recognizable shapes — shapes that make sense — emerge at last.

## HOW DEEP IS TIME?

Every journey, so the proverb goes, starts with a single step. The road to Gondwana began with a fundamental question: How deep is time?

In 1650, James Ussher, the archbishop of Armagh in Ireland, decided after years of study and calculation that the Earth and everything in it was created by God on Sunday 22 October, 4004 BC.[1] He arrived at this date after thorough and exhaustive research into ancient Latin, Greek and Hebrew texts, and of course, the Bible. As preposterous as Ussher's estimate might seem today, this kind of thinking was typical in the seventeenth century, even among some of the greatest thinkers of the day. Isaac Newton, for example, one of the most brilliant and influential scientists of all time, broadly concurred with Ussher's estimate.[2] This was a time when 90 per cent of Europe's population lived in rural areas, belief in magic and alchemy was commonplace, and at any moment a witch might be burnt at the stake.[3] Humanity was seen as central to the world's existence, and in the standard view the Earth had existed for only five days longer than humans, because that was what the Bible, the authoritative text on such matters, said about it.

To most people, the idea that Earth had a history of its own, indeed one far greater than humankind's, would have seemed utterly inconceivable. By the late 1600s, however, a new concept was taking root in the fertile soil of the Enlightenment — one that would take the best part of three centuries to become fully formed, but which would fundamentally change our understanding of our place in the cosmos.

Humanity's defining insights often disappear from plain view, as the rest of the world catches up with their originators and the ideas themselves become absorbed into the fabric of everyday reality. They become everyday truths, so self-evident as to appear almost childishly simple. The wheel, for example, is today such a commonplace, necessary and 'obvious' invention that we cannot imagine a world without it. And yet it took us tens of thousands of years to hit upon the idea. (It didn't appear until around 4000 BC, thousands of years after the invention of agriculture and pottery.)[4]

In a similar manner, we might today wonder how anyone could ever look at the world, with its soaring mountains and vast, deep oceans, and not understand that all of this has been here for a very long time. But in the seventeenth century the idea of Earth being far older than us was in no way an obvious conclusion to leap to.

Scientists of the day (although they weren't known as scientists; the word hadn't been invented yet — such seekers of knowledge were sometimes known as savants or natural philosophers) were usually religious men (and they were almost all men) who were, for the most part, seeking to reconcile the testament of Scripture with the world they saw around them. They searched old texts and manuscripts looking for historical evidence to support the teachings of Genesis, working from the basic assumption that Earth's history was also humankind's history. The Bible, of course, was seen as the most authoritative source.

But there were problems. And one glaring problem in particular: fossils. Europe was full of them. All across the continent, from outcrops along the Mediterranean coast to limestone high in the European Alps, impressions of what looked suspiciously like seashells and other living creatures were everywhere to be found eroding out of the rock.

For centuries, natural philosophers and clerics had argued about what they represented. Some thought they were 'sports of nature', planted by a mischievous God to aggravate us. Another view was that they simply grew in the earth in a kind of sympathetic vibration with the living world. And yet for perhaps as long as people had been making these speculations, there had also been those who argued what today appears so obvious to us — that fossils were the remains of once-living animals.

The Greek philosopher Xenophanes proposed in the sixth century that through time Earth was subject to alternating periods of wetness and dryness, and he saw fossil sea shells as evidence that water had once covered much of the Earth's surface.[5] In 1074 AD, the Chinese naturalist Shen Kuo observed fossils in the Taihang region and (correctly) concluded that the sea had once covered these mountains. The Italian polymath Leonardo da Vinci, digging sea shells from the Italian Alps in the early 1500s, also decided fossils looked like

*TAB. IV.*
*LAMIAE. PISCIS. CAPV T.*

*EIVSDEM LAMIAE DENTES*

*Nicolaus Steno's shark tooth fossils, 1667.*

living animals because that's exactly what they are.[6] This anticipated by more than a century the scholar Nicolaus Steno, who in 1666 compared fossilized sharks' teeth from the Italian coast with those from a recently deceased living shark to reach the same conclusion.[7]

Accepting that fossils were in fact the remains of once living creatures, however, created an enormous quandary. The problem was one of time — to be specific, the amount of it required. The Bible records the great flood of Noah's day, which was tailored to fit as an explanation. Steno postulated that the flood had played a role in depositing marine fossils in the mountains. But there were those who, perhaps privately, had their doubts. How could a sudden and brief flood, no matter how enormous, result in the burial of an animal 10, 20, 30 metres deep into solid rock?

Fossils occur primarily in sedimentary rock, which is made of silt, sand, mud, gravel and clay laid down over long periods of time on sea floors, lake or riverbeds. This type of rock forms layers, or strata. Steno realized that the layers he saw in rock formations were evidence of changing forms of deposition. In oceanic sediment, for instance, when the sea level falls, the particles raining down on the

*Sedimentary rock formation, Red Rock Canyon State Park, California.*

seabed change from fine mud to coarser sand as the distance to the shore is decreased, and so the rock this sediment forms changes too — from dark, smooth mudstone to a rougher, whiter, shallow-water sandstone. Where there has been a change from lake to river deposition, the nature of the sediment also changes, from fine-grained mud to heavier gravels. Sedimentary rock, Steno realized, had originally been laid down in horizontal beds, and in places where these beds were not horizontal it was because movement in the Earth had tilted and deformed the layers.

These insights, radical at the time, became known as Steno's Law of Superposition. Superposition was a revolutionary concept. For one thing, it implied that sedimentary rocks were not just static hunks of material that had always been so — they were in fact evidence of slowly forming rock. They were sequential, a tape recorder of time. And, as was clear to anyone who paid attention, such a sequence of events could only have occurred over a very long period. The young Earth of Ussher had been dealt a heavy blow.

By the late seventeenth century, the scientific revolution was in full swing and in the coffee houses of Paris, Edinburgh and London radical new ideas about the nature of the Earth were fermenting.

The Royal Society was established in London in 1660, bringing together some of the greatest minds of the day. In 1687, Isaac Newton published his *Philosophia Naturalis Principia Mathematica*, which laid out his laws of universal gravitation and was perhaps the single most important scientific document in all of human history. Around the same time the polymath Robert Hooke, England's answer to Leonardo, was observing microorganisms through a device he had just invented, a microscope.

Among the brightest lights of the Enlightenment era was Georges-Louis Leclerc, later known as Comte de Buffon, a wealthy natural philosopher who in 1739 established himself in Paris to devote his life to the study of the natural world.[8] Buffon was a gifted communicator and one of the most-read authors of his day. His enduring legacy would be to compile just about everything then known about the natural world into an enormous 36-volume treatise entitled *Histoire Naturelle, Générale et Particulière*. Like many of his contemporaries, Buffon was unwilling to accept that Earth's history was just a few thousand years, and he set about trying to establish how old it might really be.[9]

Buffon's theory, shared by Isaac Newton and others, was that the Earth had been formed when a comet sheared off a piece of the sun. The planet had therefore once been a fireball of molten rock. Since its formation it had been steadily cooling. If you could work out the rate of cooling, Buffon reasoned, you could estimate the age of the Earth.

To this end, Buffon began melting iron balls in his foundry and observing the time it took for them to cool. By extrapolating from these experiments, he eventually arrived at an answer. The Earth, he proposed, was 75,000 years old (privately he felt it was much, much older). This was radical thinking, and it was perhaps inevitable that his claims would raise the ire of local religious authorities, who demanded a retraction. The politically astute Buffon acquiesced and published one. However, he continued to print his books without changing a thing.

Buffon was also fascinated by the distribution of different plant and animal species around the world, and how they had come to be there — the science we now call biogeography. He was particularly intrigued by the presence of what appeared to be elephant bones in the Arctic regions. Buffon believed that as the Earth cooled, species which had originally been formed (by God) in these regions had migrated from these centres towards the warmer equatorial regions. Elephants had migrated south into Africa and Asia, where they are found today. That elephants in these places appeared to be quite different to the remains of their ancestors in Siberia, he explained by a complex theory of gradual change. Animals, he believed, grew according to an internal 'mould', which was in turn influenced by organic 'molecules' in the environment the animals lived in. His theory held that as these animals migrated southwards into new regions the particles changed, and so did the animals' forms.[10]

Buffon's ideas were typical of his time — 'grand theories' that sought to explain the workings of the world in big, tidy packages. While most of his theories would later dissolve under the harsher light of modern scientific enquiry, his writing laid the groundwork for much of what was to follow.

Like most of his contemporaries, Buffon was not a field scientist as we think of one today. He cooked up his theories in a comfortable Parisian study, relying on others to bring him his rocks, and rarely venturing into the landscape he aimed to understand. In order to really get to grips with the nature of the Earth we walk on would require a new approach — it was time for science to get its hands dirty.

A band of soft limestone stretches through the southwest of England in the region known as the Cotswolds, emerging above green fields, villages and hedgerows. Today, motorways dissect the limestone, traffic grinding over and through England's green and pleasant land. And yet, just off the motorway, the traveller may still step back in time to wander through quaint little villages that dot the region, places with oh-so-English names like Crickdale, Moreton-on-Marsh and Chipping Norton, many of them built from the limestone hundreds of years ago.

It was in one such village, Churchill, that William Smith was born to a blacksmith and his wife in 1769. Smith received little formal education but nurtured an incisive intellect and a keen observational eye for the natural world. His childhood days were often filled with wandering the Cotswolds, collecting and obsessing over the fossils of urchins and other sea creatures he found among the limestone.[11]

By now, the industrial revolution was in full swing and demand for coal to drive the steam engines and factories of Europe was soaring. As a young man, Smith landed an apprenticeship with a local surveyor who employed him to work in coal mines in the south of England, where miners daily lowered themselves on ropes and ladders into the depths of the earth to dig out the black rock that would put bread on their tables, blacken their lungs and shorten their lives. Venturing into these mines gave Smith the opportunity to observe the rock strata at close quarters. He came to know the layers intimately, and developed a keen sense of where in the pile coal was to be found. He also paid close attention to those strata that held fossils, which, he noted, were usually in close proximity to the coal-bearing seams.

Smith was hired to survey a canal system through the countryside, a more efficient

*William Smith's geological map of Britain.*

means of transporting coal from the mines to markets in London. As engineers cut into the rock, Smith was able to observe rock strata at different places along the route. He came to realize that strata he observed outcropping in one valley would often crop up again in neighbouring valleys and hills. He began to understand how these strata were bunched like pages in a crumpled-up book, right through the country.

He reached the conclusion that one ought to be able to map these strata, not just in local regions but all across Britain. Many of these rock units contained fossils, and Smith also realized that the strata could be identified by the types of fossils they contained. Following

Steno's Law of Superposition, he knew that younger rocks overlaid older rocks, and the fossils therefore followed what Smith labelled 'the Principle of Faunal Succession'.

Over many years Smith travelled the country, carefully recording his observations. By 1802, after decades of work, he had created a map that brought it all together, showing clearly how Britain was built from successions of rock, bunched and folded across the landscape. It was a monumental achievement. Smith had produced the first ever geological map — not only a highly accurate and useful tool for miners and engineers of the day, but a work of art in and of itself.

Smith's achievements were not immediately recognized. This was an era when the scientific establishment was made up of landed gentry, men prone to looking down their noses at someone like Smith, a lowly surveyor. He ran into financial difficulties and, for a spell, wound up in a debtor's prison. Late in his life, however, the Geological Society of London finally awarded him the fame he deserved, bestowing upon him its prestigious Wollaston Medal. The society's president, Adam Sedgwick, referred to Smith as 'the father of English geology'.

Smith's interests in mapping England's geology, however, were primarily commercial. He saw much effort and expense being wasted digging for coal in places where he knew, from his own observations, there was none to be found, and he sought to remedy this. It would fall to others to read the bigger story of Faunal Succession and to understand what it could tell us about the history of the world itself.

At the same time as Smith was mapping England, farmer-turned-scholar James Hutton was also travelling the countryside of his native Scotland, taking a deep academic interest in the rock layers he saw outcropping everywhere he went. His experiences as a farmer had given him a keen interest in the soils and rocks that underlay Scotland. He had 'become very fond of studying the surface of the earth, and was looking with anxious curiosity into every pit or ditch or bed of a river that fell in his way'.[12]

Among many other things, Hutton observed how strata in outcrops had been upended and subsequently ground down, with new layers added on top of these older ones. Hutton realized that such monumental processes could not happen in a short period of time. He envisaged the continents being slowly ground down and eroded away, only to form new continents at the bottom of the sea. The world, he came to understand, was subject to slow-moving but immensely powerful forces occurring over vast amounts of time — an eternity, he believed. These Earth processes, he wrote, had 'no vestige of a beginning and no prospect of an end'.

Hutton's fellow Scotsman and geologist Charles Lyell concurred with Hutton, arguing that all of Earth's geological features could be explained by processes we see occurring in front of us: a concept that became known as uniformitarianism. Watch a mountain stream gently eroding sand and gravel from its banks. Imagine that process going on for a long, long time and you can begin to understand how, grain by grain, that tiny stream could eventually grind a mountain range down to sea level. Uniformitarianism is a very boring word for a spectacularly mind-blowing concept — the immense age of the Earth, revealed in rock. Lyell collected these ideas in a book he called *Principles of Geology, Being an attempt to explain the Former changes of the Earth's Surface, by References to Causes now in Operation*, which would go on to become a pivotal text in the rapidly emerging science of geology.

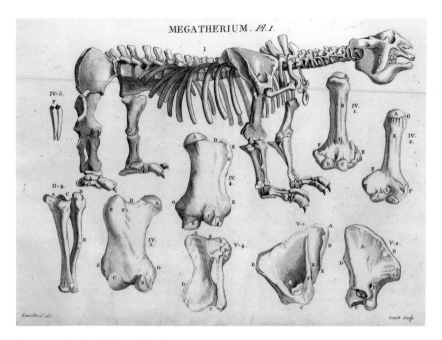

*Georges Cuvier's* Megatherium.

By the late eighteenth century, the movement of intellectual thought that typified the Enlightenment in Europe was well and truly gathering steam. In France, dissatisfaction with the privileged rule of the centuries-old ancien régime exploded into revolution in 1789. The King and Queen were hauled from their extravagant place and executed in 1793. After the revolutionary bloodletting which followed, Napoléon Bonaparte staged a coup d'état, declaring himself Emperor in 1799.

In the wake of the revolution Paris emerged as a hotbed of intellectual study. The Revolution established a new order based on Enlightenment principles, and great minds flocked to the city to take advantage of the opportunities this offered.

Among them was Georges Cuvier, who at the age of 26 became a junior scholar at the Museum of Natural History. Cuvier was a pioneer in comparative anatomy, the field of science based on the fundamental idea that all animals share the same basic body plan. A gorilla and a pelican, for example, might be very different beasts but each has the same basic array of bones and organs, albeit shaped and structured differently, and used for different purposes. The powerful arm bones the gorilla uses to haul foliage from the forest are, in a pelican, a delicate wing structured for graceful flight.

By comparing the bones of different animals side by side, it is possible to see how closely related they are. Headstrong and opinionated, Cuvier prided himself on his ability to identify animals based on just a single bone. He made his mark early on, after being sent the engravings of a set of giant bones that had been found in South America. Cuvier analyzed these alongside bones in the museum's enormous collection and declared, correctly, that this fearsome animal's closest relative was the humble tree sloth. He named the creature *Megatherium*, or 'huge beast'. What was especially interesting about this monster was that no living example had ever been reported, which, given the Spanish had by now been stomping around South America for over three centuries, implied it no longer existed. At the time, it was widely considered impossible for a species to go extinct. Every creature was one of God's perfect creations, part of a great, unbreakable chain of being, so the thinking went. How could He possibly allow one to vanish from the face of the Earth?[13]

Cuvier disagreed. He had also gotten his hands on a box of large 'elephant' bones that had been recovered by French soldiers in the wild regions of North America, freighted down the Mississippi River and sent to Paris. The beast, dubbed 'the Ohio animal', presented a quandary. It was clearly not the bones of any animal known, resulting in two possibilities. Either the Ohio animal was still out there to be found in the poorly explored American wilds, or the skeleton belonged to a species that was now extinct.

Like Buffon, Cuvier had also examined bones from Siberia, which again superficially resembled elephants but were much larger than any elephants known. When Cuvier examined them, he found them quite different to living elephants. In 1796 he announced that neither these nor the bones of the Ohio animal belonged to living animals but were the remains of creatures that had entirely vanished from the face of the Earth.

He gave the name *Mastodon* to the Ohio animal, while the animal from Siberia he called a mammoth, names we still use for these extinct Pleistocene creatures today.

Cuvier came to believe the Earth had at numerous times been subject to sudden and violent 'catastrophes'. A religious man, although not rigorously so, he thought that Noah's flood had been just one of these catastrophes. It was these periodic upheavals that had made species like the Ohio animal disappear *en masse*. These ideas were widely challenged. While Lyell, for example, was prepared to accept the truth of extinctions, he could not reconcile with Cuvier's sudden, overwhelming catastrophes — they didn't fit with his concept of achingly slow, uniformitarian Earth processes.

This conflict, between Lyell's slow uniformitarianism and Cuvier's catastrophes, was central to early nineteenth-century science. In time, both would be proven right, and wrong, in equal measure.

Besides building a new nation from the ashes of the old, there were more mundane tasks at hand in Napoleonic France. One of these was to ensure a national supply of ceramics. The government-owned ceramics factory Manufacture Nationale de Sèvres was renowned throughout Europe for its high quality, ornate work, and Napoléon was keen to ensure its continued success, as a symbol of the prosperity of the new nation.

The man chosen to direct the factory was Alexandre Brongniart, a young chemist, zoologist and mineralogist whose managerial and business acumen,

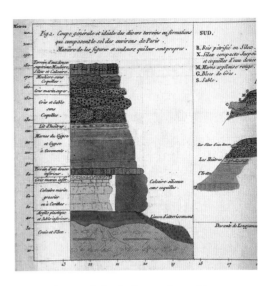

*Brongniart and Cuvier's stratigraphic profile of the Paris Basin.*

combined with his scientific knowledge, made him the man for the job.[14] Brongniart's first love, however, was natural history. Among other things, he had contributed ground-breaking research into the taxonomy of reptiles. His day-to-day role at the ceramics factory was not just managerial; he was also required to apply his understanding of geology to source the clays necessary for the manufacture of ceramics from the area around Paris.

Brongniart's explorations took him deep into the geology of the Paris region, where he was able to observe the complex layering of rock strata around the city and see first-hand the fossils these layers held.

Brongniart had previously travelled to London, where he had, in all probability, seen an early draft of William Smith's map. He was well aware, therefore, of the power of fossils to demarcate rock units.

Brongniart was joined in these explorations by his friend Georges Cuvier, and together they painstakingly explored and mapped the Paris area. Through meticulous field work they ascertained that Paris sat in the middle of a big depression formed in the underlying chalk (the same unit of rock that is exposed as the famous white cliffs of Dover). This chalk layer was overlain by a sequence of sedimentary deposits, some of which contained the fossils of marine invertebrates, including extinct sea creatures. There was also one layer of gypsum that contained numerous mammal bones, again of species that were not to be seen living at the time.

Unlike Smith, who primarily viewed fossils as convenient markers to delineate different rock types, Cuvier and Brongniart recognized fossils were crucial pieces of evidence that could tell a story through time — 'nature's documents', as Cuvier called them. The fossils of the basin showed that over a long period of time, sediments had been laid down in both freshwater and saltwater environments, and that these modes of deposition had alternated back and forth through time.

In 1811, Cuvier and Brongniart produced a map similar to Smith's, as well as a detailed cross section, which clearly laid it all out for the world to see. In the words of geological historian Martin Rudwick:

> *Cuvier and Brongniart presented a geohistory of the Paris region that was as complex and ... unpredictable as the bewildering twists and turns, the war and peace, the sudden coup d'état and quiet interludes of the Revolutionary and Napoleonic politics they had both lived through in the past two decades.*[15]

In a broader sense, though, what the pair had done was set the agenda for the next two centuries of geological enquiry; they had formulated a method of using strata, and the fossils they contained, to tell big stories about the Earth's history.

The old edifice of a young, stable Earth was crumbling. No longer was the Earth a stage provided for humanity's benefit alone. Suddenly, ours was just a bit part in a planet-scaled drama far greater than we could ever imagine.

Outside the University of Otago's geology department, a series of rock displays are presented on stone plinths beside the building's three main entrances. The oldest is a big chunk of chlorite schist, a rock formed from sediments laid down in the sea off the coast of what is now Australia, around 160 million years ago. It was buried 15 kilometres (9½ miles) in the Earth's crust, subjected to huge temperatures and pressure, and baked to form the foliated rock, glistening with mica crystals, so prominent in the hill country of our region.

Next to this is a granite pegmatite, a once-molten rock extruded in the Earth 40 million years later, cooling slowly enough to allow massive pink and white feldspar crystals to form. The next two blocks hold marine sediments — a 60-million-year-old sandstone and a much younger limestone, both remnants of ancient seas in which weird shark-toothed dolphins and giant penguins once swam. The final rock samples are of hexagonal basalt columns that were formed when the volcano that the city of Dunedin is built on was erupting from the Earth around 10 million years ago.

These rocks, in very broad terms, tell the basics of New Zealand's geological story. It is just one small part of a much, much bigger story — the story of the supercontinent Gondwana, its formation and subsequent break-up. It's a story that connects my country with Australia, Africa, the Arabian Peninsula, India, South America, Papua New Guinea and Antarctica.

These rocks, in all their gnarled and twisted glory, speak of a rocky surface in constant motion and of the life we live on shaky ground here, with earthquakes and volcanic eruptions a constant threat. The Earth beneath our feet is anything but stable.

What's really astonishing is that this understanding is one that only fully came into focus in the last 50 years, within the careers of some of the older professors still working in this very building.

In the early 1800s, when Cuvier and Brongniart were hard at work amid the strata of the Paris Basin, there was still a long way to go. Many more minds would need to be bent to the task before we would truly begin to understand the fluid, violent nature of the ground we stand on. And central to it all would be the unassuming fossil leaf that sits in the corner cabinet of the University's geology museum. By the time all of this was over, *Glossopteris* spp. would have taken on a significance that was literally earth-shaking in scale.

# 750 MILLION YEARS AGO

Her name, in a nameless time, is Rodinia, which, in the words of a language that will not exist for another 750 million years, means 'to give birth'. She is, in a sense, mother to us all. You might call her an island, being a landmass surrounded by sea. Except this is a mega-island. A supercontinent composed of all the land on this planet above sea level — something like 150 million square kilometres (58 million square miles) of it, all pushed into one giant heap.[16]

Rodinia's shores are pounded by a vast sea — by default, all of this world's oceans in one, called Mirovia. Mirovian storms generate waves that, with an entire globe to run amok on, gather huge bulk and height, slamming into Rodinia's long shores with unimaginable force, gouging her flanks into precipitous cliffs that stream with water.

The land itself is dead. This planet's atmosphere is extremely low in oxygen; if we were to try to breathe here, we would quickly suffocate. Complex life forms have yet to evolve on this world, and even if they had, the lack of an ozone layer would make life on Rodinia intolerable, as solar radiation bakes the landscape and poisons the light.

The continent is frequently raked by storms that wash massive amounts of sediment into the surrounding ocean, staining it a rust-red colour. These seas also appear dead, but on a microscopic level Mirovia pulses with single-celled organisms — bacteria — tiny beings that have been around since the dawn of life itself some 3.5 billion years earlier. They feed on the sun's rays, converting light energy to chemical energy. As a waste product of this biological alchemy they produce oxygen which, molecule by molecule, is slowly filling up the Earth's atmosphere.

For 500 million years, Rodinia has sat stable in her position on the southerly edge of the globe, her low-density basement rock holding her 'afloat' on the planet's mantle. At the edges of the continent slabs of dense, heavy oceanic crust slip under Rodinia, to be reabsorbed in the churning mantle below.

For the past few million years, in one corner of Rodinia, some of that oceanic crust has been getting stuck and huge piles of it are now wedged against the side of the continent. For a long time that material has piled up, the immense pressure growing and growing, until finally something gives. The blockage falls away and, over millions of years, the entire mass plunges deep into the mantle, delivering a rapid overdose of combustible material to the inferno below. In the fury that follows, a huge plume of hot convection billows up deep beneath Rodinia. It pushes up against the bottom of the continent like a helium-filled balloon.

Out in the Rodinian desert, the ground starts to rumble. An explosion tears the boiling air, and the Earth gouges open, spilling its guts across the landscape in the form of hot magma. Great plumes of black smoke and ash fill the sky, blocking out the sun and pumping carbon dioxide into the atmosphere.

Rodinia is thinning, weakening and stretching. The motherland is breaking apart.

# 2

# THE DARKNESS IN THE MOUNTAIN

For tens of thousands of years, the ancestors of the Awabakal people made their home along Australia's southeastern seaboard. Their traditional name for the area is Nik-kin-ba, which means 'the place of coal'.

Long ago, according to their oral tradition, the world was plunged into a darkness that blocked the sun. The darkness came from a hole in a mountain. People came from all around to cover the ground with rocks and sand, fearing that the ever-burning fires deep in the ground would release the darkness again.

After the darkness was covered over, generations of people walked on the ground, pressing the darkness and the flames together under the earth to become nikkin, or coal. Now, whenever that coal is burned, the legends say, the spirit of that ancient fire is released.[1]

In the 1700s European settlers arrived in Australia to expand Britain's colonial possessions. The name they imposed on this region was New South Wales. Port Jackson, the settlement that would become Sydney, began as a penal colony to which Britain shipped her convicts.

For most of these unfortunate souls, transportation to Port Jackson was a life sentence, as there was often very little hope of returning home. Among them were William Bryant, a smuggler, and his wife, Mary, who found themselves cast upon these bitter shores in 1791. Conditions in the colony were brutal, with starvation and violence commonplace, so the Bryants made the decision to escape. They stole a small boat, and along with a group of fellow convicts sailed it out of Port Jackson. Their intended destination — Indonesia, far to the north.

Two days out from Port Jackson, they pulled into a sheltered place called Glenrock, where they hauled up for the night. Here, they cooked their evening meal over hunks of coal they found littered along the beach. In doing so they became the first Europeans to 'discover' coal in New South Wales. The next day, they continued their voyage. After 69 days and 5236 kilometres (3253 miles), having completed one of the longest and most audacious small-boat voyages of all time, they reached Timor, only to be recaptured and returned to the penal colony.[2] Coal mining would go on to become one of New South Wales', and Australia's, most important industries.

The oft-told story of the Bryants' 'discovery' of coal diminishes the thousands of years of prior claim of the Awabakal people. But then, this was an age of British 'discovery'. All around the Empire, colonial agents were busy claiming for King and country the treasures of the old Earth. Central to the global colonial expansion was the search for coal. And in seeking it, scientists would open a window into the deep past. It was in the coal mines of the British Empire that they would first glimpse Gondwana.

Coal is formed in ancient swamps or estuaries from the remains of decaying vegetation. It forms where forests produce more matter than can be decomposed in the air by bacteria and fungi, leading to a build-up of organic material in the ground.

As more and more plant matter builds up, it is buried far beyond the reach of oxygen-loving bacteria that would otherwise consume it, until, over millions of years, it is compressed and heated to form a brown rock called lignite. Over time, the rock will lose water, thus becoming more carbon-rich, forming bituminous coal and finally anthracite, the highest grade coal of all.[3]

We humans never found much use for coal until the late eighteenth century when the invention of the steam engine, the advent of factory production and an increasing dependence on iron and steel created a huge need for energy. Coal by the train load was needed to power the factories, fuel the smelters and drive the steam engines of the industrial revolution.[4]

There is no definite word in the classical Indian Sanskrit language for coal. That is probably because in pre-colonial times much of the continent was covered in dense forests that would have provided all of the villagers' combustible needs.[5] In the 1700s India was dominated by one awesomely powerful corporation, the British East India Company, which had grown from its humble beginnings as a small monopoly engaged in the Oriental spice trade to become the primary agent of British colonialism on the subcontinent.[6] The company was, in effect, in the business of running India for the profit of its shareholders. At its peak, it ruled a fifth of the world's population, generated more revenue than all of Great Britain and even had its own private armed force, a security contingent twice as large as the British Army. In 1757 the company seized control of the state of Bengal and began draining it of its riches.

In those days, when forest still encroached on Bengal's Damodar River and the land shrieked with wildlife, the local Bauri villagers would collect coal from seams along the riverbanks to make their cooking fires.

When Suetonius Grant Heatly, a British East India Company official, travelled up the Damodar and saw Bauri farmers using coal in their villages, he knew he was onto something big. According to one report, Heatly kidnapped the local rajah and extracted from him the intelligence as to where the best coal in the area was to be found.[7] With this knowledge in hand, Heatly, along with another company man, John Sumner, applied for a permit to mine the region. In 1774 they began employing local Bauri people to dig coal out of tunnels in the ground. Over the coming decades, numerous mining ventures would spring up along the Damodar Valley.

Problems with land disputes and the ongoing issue of finding enough labour hampered operations. English miners frequently succumbed to the tropical diseases that festered in the Damodar's mosquito-plagued waters, while the river itself ran so low through the dry season that transporting the coal downriver to Kolkata was only feasible for a few months a year. While sitting in storage, the coal, along with its market value, would degrade.

And so the nascent Indian coal industry limped along in an unsatisfactory manner until 1814. By now, though, demand for coal was soaring, with global production having doubled, then doubled again since the early part of the previous century. Suddenly, the economics of mining in Bengal started to make much more sense.

The East India Company sponsored mining engineer, William Jones, to invigorate the industry in the region.

Jones leased a stretch of land in the Damodar Valley from the rani, or princess, of the area; a place named in her honour, Raniganj. Here, industrial-scale coal production in Bengal began in earnest. In time, Raniganj would become the powerhouse of India's coal industry, and a place synonymous with both profit and pain.

# FINDING *GLOSSOPTERIS*

The development of coal mining around the globe had a side benefit for science — fossils found as part of the exploration work were being regularly sent to institutes of learning in Europe, where scholars had an opportunity to study them.[8]

In the early 1800s, paleobotany, the study of fossil plants, was still in its infancy. It would fall to a man living in a revolutionary time, and with a heavy-duty scientific pedigree behind him, to really kickstart the science.

Alexandre Brongniart's son, Adolphe-Théodore Brongniart, was born with the new century on 1 January 1801, into a nation that had cast off many of the old institutional ways of thinking and embraced a new openness to innovation and enquiry. As a youngster, Brongniart had accompanied his father on geological research trips across Europe, where he had been exposed to the mysteries and wonders of the fossil world while steadily gaining an unusually broad overview of the geology of the continent.[9] By now, geologists were becoming much more familiar with the layers of rock strata that made up continental Europe. Initially, the geological sequence had been separated into just three sections: the Primary, or basement rocks; the Secondary strata, which overlaid the Primary and were often full of fossils; and the Tertiary rocks, the youngest sediments.

The pressing task in the early part of the nineteenth century was to break these large units up into smaller parts. Over two decades, geologists identified and named nearly all the divisions of the geological timescale that we are familiar with today.

The Tertiary was divided into the Pliocene, Miocene and Eocene epochs by Charles Lyell. German paleontologist Heinrich Ernst Beyrich later proposed a further epoch, the Oligocene.

At the top of the Secondary pile sat the Cretaceous strata, which take their name from the Latin 'creta', meaning chalk. The Cretaceous was named for the distinctive layers of chalk of this age that outcrop in England and France (such as the chalk deposits of the Paris Basin that Brongniart and Cuvier had studied). Below the Cretaceous came the Jurassic, named by Alexandre Brongniart for the Jura Mountains on the border of France and Switzerland, where rocks of this age were well exposed. Underlying the Jurassic was the Triassic, given its name by the threefold layering that this part of the record typically exhibited throughout Europe. Further down in the pile, the Permian period was named for rock outcrops in the Perm region of Russia, and the Carboniferous below it for the massive amounts of coal it held (almost all the coal in Europe is mined from Carboniferous rocks). Beneath the Carboniferous lay the Devonian, which was named after Devon in England, the Silurian and the Ordovician, each named after ancient Welsh Celtic tribes, and finally, the Cambrian, which borrows its title from the ancient name for Wales.

As geologists became more familiar with Europe's fossils and how they related to each other in chronological sequence, striking patterns in the development of life were emerging. Mammals, the fossil record showed, had been prevalent throughout the Tertiary record, just as they are today. But when you went deeper in time, down to the Cretaceous, Jurassic or Triassic rocks, the fossils revealed a very different world indeed.

Cuvier examined the bones of a huge animal found in Cretaceous sediments near Amsterdam, a fossil known as the 'Maastricht animal'.[10] Sensationally, he declared the Maastricht animal to be a giant marine lizard, an extinct animal he named a mosasaur. Other such discoveries, including the discovery of the first dinosaur fossils by Gideon Mantell and others in the 1820s, showed the later part of the Secondary period had been an age of giant reptiles.

The older parts of the Secondary record revealed an even stranger time, the Devonian period, when there were no reptiles and mammals at all, but lots and lots of very weird fish. Rocks at the bottom of the heap, in the Cambrian, Ordovician and Silurian periods, recorded ancient seas inhabited by bizarre jawless fish similar to lampreys or hagfish, and scuttling sea floor creatures known as trilobites.

It was becoming clear, then, that in geological time, life on Earth was not fixed — there was a clear, unidirectional development of animals. In the early parts of the fossil record, animals appeared 'primitive' but as time progressed 'higher' forms of life — the reptiles, birds and mammals — appeared.

Adolphe-Théodore Brongniart would turn this kind of revolutionary understanding to plants. Fossil plants, he realized, could provide a glimpse into past climate. Examining the Carboniferous-age coal deposits in Europe led Brongniart to believe that the forests which formed them had grown in much warmer times than currently persisted on the continent. He proposed that these great vanished 'coal forests', which had formed such vast and economically important seams throughout the Northern Hemisphere, had existed in a time of high 'carbonic acid', or carbon dioxide. 'Of these several distinct associations of vegetables which have successively flourished on our globe,' he wrote, 'none is so worthy of our attention, as that which appears to have been first developed (Carboniferous) and which, during a long space of time, seems to have covered with thick forests all those parts of the Earth which were above the level of the waters, and the remains of which, piled upon one another, have formed those frequently thick and numerous beds of coal, which are preserved relics of the primitive forests that preceded, by many ages, the existence of man.'[11]

Through photosynthesis, plants draw in carbon dioxide and release oxygen as a waste product. As concentrations of carbon dioxide declined with the proliferation of new forms of plants, Brongniart thought, the atmosphere had become more oxygen-enriched. It was this increase in oxygen that had allowed the higher forms of animals to come into existence.

Between 1828 and 1838, Brongniart compiled all of the world's current knowledge of fossil plants into his *Histoire de Végétaux Fossiles*, a book which would establish him as a pioneer of paleobotany. In this book, Brongniart laid out a broad concept of how the Earth's flora had changed through time, a story he divided into four parts. The oldest plant fossils were those that formed the coal strata. These plants had no flowers and despite being, in many cases, large trees, seemed to be most closely related to ferns, liverworts and mosses. After that, the fossil record was dominated by cycads, then conifers — the ancestors of our modern-day pines, spruces, larches and redwoods; plants without flowers but with cones that partially enclose the seeds. Finally, in the late Secondary and early Tertiary periods came the flowering plants: the group known as angiosperms, which dominate most parts of the world today.[12]

Included in Brongniart's *Histoire* were fossil plant specimens from all over the world, including those filtering back to London and Paris from the colonies. Among them were fossils hacked from the sunbaked, coal-filthy Indian earth at Raniganj. The most common of these were the fossilized leaves of a plant that seemed to have existed throughout the period that India's coal was forming, then suddenly vanished. These fossils were of a distinctive, long, paddle-shaped leaf marked with a network of veins flowing across its surface like tributaries of a river, towards a central midrib that finally ran towards a gently rounded tip. The leaves looked, to Brongniart, like a tongue. To his surprise, the Indian fossils bore a striking similarity to fossils he'd also examined from coal deposits in Australia.

'The leaf of this plant,' he noted in his description of the fossils, 'appears to be very common in the coal mines of New Holland [Australia]. These are the same fossils that some travellers had regarded as *Eucalyptus* leaves; but a closer examination shows that the arrangement of their veins is quite that of ferns, and bears no resemblance to that of *Eucalyptus*.'

In 1828, Brongniart formally described the plant using two key specimens. One came from a coal mine along the Hawkesbury River in Australia's New South Wales. This one, he called 'Variation a, *Australasica*'. The other came from the coal mines of Raniganj, and this he described as 'Variation b, *Indica*'.[13]

'The unique sample from India,' Brongniart wrote, 'differs from Var. a only by its larger size, and its more regularly lanceolate shape, which seems to end in a more acute point.' Despite these minor differences, Brongniart was convinced that Variation a and

*Glossopteris browniana.*

Variation b belonged to the same species. As the first taxonomist to describe them, it fell to Brongniart to give them a name. Their appearance, he thought, offered one. In Ancient Greek, the word for tongue is *glossā*, while the word for fern is *pteris*. He put these descriptive words together to create a new genus: *Glossopteris*, the tongue fern. He then gave them the species name *browniana* after his renowned colleague Robert Brown.

And so it was that *Glossopteris browniana*, an attractive but otherwise unremarkable fossil, entered the annals of scientific literature for the first time, as just one of hundreds of ancient plants being described at the time. Brongniart surely could not have foreseen the immense significance it would eventually take on.

To appreciate the story *Glossopteris* had to tell, and to visualize the mysterious place it could lead us to — Gondwana — would take a global perspective that was still decades away.

# 'A ROCK ABOUNDING WITH SHELLS'

The Falkland Islands, or Islas Malvinas as the Argentinian people know them, lie 480 kilometres (300 miles) off the coast of Argentina. In 1833, the islands were inhabited by millions of penguins, some half-wild cattle and a handful of tough, Spanish-speaking gauchos. On 1 March that year, a ship called the *Beagle* dropped anchor in Berkley Sound on East Falkland. Aboard was the 24-year-old naturalist Charles Darwin, who was eighteen months into his famous journey around the world.

Before arriving in the Falklands, Darwin had explored the rainforests of Brazil and the plains and peaks of Patagonia. His constant companion on the voyage had been his treasured copy of Charles Lyell's recently published *Principles of Geology*. As well as studying the living world, Darwin was also deeply interested in fossils.

Perhaps having been spoiled by the riches of the tropical rainforest and the majesty of the Andes, Darwin's first impressions of the Falkland Islands were not entirely favourable. 'The land is low and undulating with stony peaks and bare ridge,' he wrote. 'The whole landscape ... has an air of extreme desolation.'

Darwin went ashore and, with a gaucho guide, walked across to the small settlement of Port Louis, where he made a discovery that would greatly improve his impression of the place. 'The whole aspect of the Falkland Islands, were ... changed to my eyes from that walk,' he recorded, 'for I found a rock abounding with shells; and these of the most interesting age.'[14]

Darwin had stumbled on the fossilized remains of brachiopods, an ancient form of marine creature that superficially resembles a clam. The rocks they were encased in were of Devonian age. Darwin brought the fossils back to the *Beagle,* where they were carefully stowed alongside the many other specimens he had collected already.

The *Beagle* sailed to Argentina and Chile, where Darwin would be fascinated by seashells he found embedded in rocks high above sea level. There, he experienced an earthquake and witnessed how mussel beds were uplifted from the sea by it. In such a way, he realized, a mountain range as grand as the Andes could be raised, given enough time.[15] This was Lyell's uniformitarianism — slow Earth processes — occurring right in front of Darwin's eyes. The limits of time were expanding away from Darwin into the vast, unknowable distance.

Over the next two years, the *Beagle* would circumnavigate the globe, stopping in at the Galápagos Islands, New Zealand and Australia before returning home to England in 1836.

In 1856, Darwin's Falkland Islands brachiopod fossils came to the attention of pioneering paleontologists John Salter and Daniel Sharpe, who were well versed in the fossil-bearing rocks of Devonian age from South Africa. To their astonishment, Darwin's fossils immediately recalled brachiopods they knew so well from the Cape district, 6300 kilometres (3914 miles) away from the Falkland Islands. ' ... The only locality where any of these South African species have previously been found,' they wrote, 'is in the Falkland Islands; and it is very remarkable that, of the nine species brought from those islands by Mr Darwin ... five are contained in the collection from the Cape.'[16] Like Brongniart's *Glossopteris* leaves, Darwin's brachiopod fossils were speaking to us, telling us a story of ancient connections between landmasses. But what kind of connection?

In the 1840s, Andrew Geddes Bain knew the wild veld of South Africa better than just about any European alive. He was already 40 years of age when a copy of Lyell's *Principles of Geology* landed in his hands. As it had for Charles Darwin, the book would have a defining influence on his life. But unlike Cambridge-schooled Darwin, the Scottish-born Bain had little in the way of formal education. He had lived an eventful life, having travelled much of the barren interior of his adopted homeland of South Africa as a trader, saddler, soldier and surveyor.[17] He had explored new (to a European) routes through the country, and on at least two occasions risked his life in dangerous engagements with local tribes. He had even tried his hand at farming, at the same time as raising a large family. Working for the South African government, Bain was also a builder of roads, and cutting new routes through South Africa's dry, shrubby interior had opened his eyes to the country's fascinating rock stratigraphy.

It was around this time the discovery of dinosaur remains in England by Gideon Mantell, and their description by Richard Owen in 1841, had begun to reveal the lost fauna that had walked the Earth in ancient times. Inspired by these discoveries, Bain set out to search for similar fossils in the South African wilderness. He soon began turning up the fossils of large animals from the rocks of the Karoo, the dry, wild region that stretches across most of the western part of South Africa. Among them were the remains of creatures with a turtle-like beak and no teeth except for two tusks protruding from the upper jaw. Bain named these animals 'bidental'.

Within a short time, Bain had amassed an enormous collection of animal fossils, enough to fill a room he rented for the purpose in Grahamstown. A journalist who laid eyes on the collection wrote that 'lovers of paleontology may expect to have a treat equal to any

*The skull of a Permian dicynodont from South Africa.*

since the discoveries of Cuvier in the Paris Basin, and those of Dr Mantell in the weald of Sussex'. The highlight of the collection, wrote the journalist, was the spectacular skull of what Bain had dubbed the 'Blinkwater monster', an enormous beast with a chomping jaw studded with 56 incisor teeth.

Bain sent his fossils to the famous anatomist Richard Owen in London, who immediately recognized them as a hitherto undescribed group. Bain's 'bidental' he named *Dicynodon*. The 'Blinkwater monster' he described as *Pareiasaurus serridens*.

A pious man, Owen was so moved by the specimens that he described them as illustrating 'a power transcending the trammels of the scientific system'.[18] What Bain had discovered were not dinosaurs, but the remains of giant therapsids and early reptiles that pre-dated the dinosaurs by tens of millions of years and which were, at the time, unknown to science.

Despite being untrained as a geologist, Bain had almost single-handedly brought to the world's attention the incredible Permo-Triassic fossil fauna of the Karoo. In his wake, many more would venture into this stark wilderness, unearthing a flood of spectacular fossils. But perhaps Bain's greatest legacy was to produce the first ever

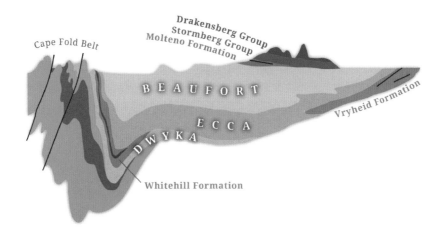

*Cross-section of the Karoo Supergroup, South Africa.*

geological map of South Africa. He sent this map to the Geological Society in London in 1852, who described it as 'the triumphant results of the single-handed labours and unaided research of one who, by his own perseverance and talents alone, has not only worked out so grand a geological problem, but has trained and wholly educated himself for the task.'[19]

Like William Smith in England before him, Bain's map showed in beautiful detail how the Karoo had been formed from successive layers of sediment, laid down over a huge period of time between the Carboniferous and Jurassic periods — a sequence of rocks that would become known as the 'Karoo Supergroup'. The Karoo Supergroup is a neatly organized pile, arranged like a plate of pancakes, which covers much of South Africa to a depth of up to 12 kilometres (7½ miles). At the bottom of the heap is a jumbled boulder bed that Bain named the Claystone Porphyry, but which would later become known as the Dwyka Group. The Dwyka is a legacy of ice. It's a tillite, a mess of loose, fragmented boulders and sludge bulldozed into place by the massive movement of glaciers across the land, before eventually being turned by time and pressure to rock.[20]

On top of the Dwyka lie two distinct formations of layered sedimentary rocks: the Ecca formation, named by Bain for a small town in an area where he had been cutting a military road at the time, and the Beaufort Group. These shale and sandstone layers, which are of Permian age, being 250–300 million years old, are full of animal and plant fossils. They are the legacy of ancient seas, lakes and rivers that once covered the region, and estuaries that were densely forested during the Early Permian. Both the Beaufort and Ecca groups contain masses of *Glossopteris* leaves. These *Glossopteris* leaves would soon connect the Karoo to another, very different part of the world.

# FOREST OF THE GONDS

In 1795 the British explorer J.T. Blunt ventured into the barely known central region of India, an area described by academic Pratik Chakrabarti as 'a vast trap country ... a hauntingly beautiful area of deep forests, ravines, hills and rivulets.'[21] Blunt's Hindu guide warned him it was an abode of 'wild beasts, demons and the savage Goands'. The latter reference was to the Gond people, indigenous Indians who had inhabited this part of the subcontinent since time immemorial.[22] This land, Blunt was told, belonged to the 'the ancient Rajahs of Goandwannah'. (Gondwana, in the local language, means 'Forest of the Gonds'.)

Gondwana was a region that inspired fear and wonder. William Sleeman, British magistrate of the central provinces, visited the area in the 1830s and noted that, according to local myths, the Pạndavas, the five central characters of the great Hindu religious epic the *Mahabharata*, had rested on the banks of the Narmada river. 'Every

*Narmada River, India.*

fantastic appearance of the rocks,' he wrote, 'caused by those great convulsions of nature which have so disturbed the crust of the globe, or by the slow and silent working of the waters, is attributed to the god-like power of those great heroes of Indian romance.'[23]

British explorers would venture deep into this semi-mythical place in search of minerals, particularly coal and the famous black soil of the region, known as regur, which was invaluable for cotton production. Fossils, they discovered, were common there. In the 1850s the Scottish geologist Stephen Hislop described the region as an 'ancient lake'. 'On the shores of this lake,' he wrote, 'earthworms crawled, and small reptiles crept over the soft mud. In its pools sported flocks of little Entomostracans [small crustaceans] mingled with which were Ganoid fishes and Labyrinthodonts [amphibians].' Plants, too, were common in the sediments. 'Low-growing plants with grooved and jointed stems inhabited the marshes,' he recorded, 'conifers and ... palms raised their heads aloft.[24]

The rocks that underlie the Gondwana region form a succession of sedimentary deposits, underlain by a jumbled bed of boulders named the Talcher Formation. Henry and William Blanford, British brothers working for the British Geological Survey, examined the Talcher in the 1850s and theorized it had been formed under glacial conditions, something that seemed highly unlikely in the sweltering heat of tropical India. Their ideas were scoffed at for another two decades until striations, grooves gouged by the movement of a glacier across bare rock, were discovered and their theory was confirmed.[25]

Against all odds, it was apparent that this part of India had, in exactly the same way as South Africa's Karoo, once been buried in ice.

The rocks above the Talcher form four units: the Kaharbri and Barakar Formations, the Barren Measures and the coal-bearing Raniganj Formation, from which so much of India's mineral wealth would one day derive. The whole sequence covers a vast swathe of India — around 50,000 square kilometres, (20,000 square miles) and is up to 7 kilometres (4.3 miles) deep.[26] These units are a succession of deposits laid down by ancient lakes and rivers. All of them contain a profusion of *Glossopteris* fossils. In their fossil content, they closely reflect the layers of the Karoo succession, 10,000 kilometres (6215 miles) away across the Indian Ocean.

The Irish geologist Henry Medlicott called the entire sequence of rocks the 'Gondwana' sequence.[27] With its connotations of mystery and wonder, it is perhaps no surprise that the name stuck. In time, 'Gondwana' would become more than just a title relevant to this economically important part of the Indian subcontinent. It would become synonymous with all that is vast and mysterious about Earth's deep history.

'Gondwana' would become a journey into the unknown.

# 3

# GREAT SOUTHERN LAND

*Terra Australis Incognita*: the great, unknown southern land. For generations we have sought it, as if somewhere in our being there lies an ancestral memory that draws us to it, knowing, or believing, that it is out there to be found. Perhaps by writing this book — by going looking for the lost supercontinent Gondwana — I, too, am seeking it.

The existence of a great southern land was theorized in Roman maps from the second century, and ever since Ferdinand Magellan rounded Cape Horn to enter the Pacific in 1520, European explorers have sought to fill the vast blank space at the bottom of the world's maps with land. In the 1700s, it was widely believed such a huge continent must exist to balance out the world's landmasses.

The epic voyages of Abel Tasman and James Cook showed that south of Africa, South America, Australia and the outlying fragments of New Zealand, there was no huge landmass filling up the southern globe — just a vast, dangerous ocean. And beyond that, the real *Terra Australis Incognita* — Antarctica, looming out of the mist and snow, bristling with ice.

It was to Antarctica that European explorers increasingly turned their attention in the nineteenth century. Cook had swung below the Antarctic circle three times between 1773 and 1774, while in 1820 the Russians sent the spectacularly named Captain Fabian Gottlieb Thaddeus von Bellingshausen to explore these dangerous waters. Bellingshausen got as far south as Queen Maud Land, where he and his crew were probably the first people to lay eyes on the icy continent.[1]

In 1839, the British explorer James Clark Ross departed for the ice from Hobart, Tasmania, as leader of an expedition aboard two famous ships — *Erebus* and *Terror*. Over a period of three years, between 1839 and 1843, Ross and his crew explored much of the Southern Ocean, heading deep into the ice pack, where they became the first people to reach the region that would later become known as the Ross Sea.[2]

Ross had been the first to locate the Earth's north magnetic pole in the Arctic several years earlier. His primary goal on this voyage was to try to find its antipodean counterpart, the south magnetic pole, something he was unable to do. The problem was that, unlike in the Arctic where the pole lay over water, in Antarctica his way was blocked by the seemingly impenetrable snow- and ice-covered land. The youngest member of the expedition was the 23-year-old botanist Joseph Dalton Hooker, an ambitious young scientist who, in the months leading up to the expedition, had slept with a draft of Darwin's *Voyage of the Beagle* under his pillow.[3]

In between their Antarctic voyages, the ships spent time in New Zealand, Tasmania and the Falkland Islands, where Hooker had the opportunity to study the flora and fauna of these places. He collected many plant specimens, building up a broad picture of how plants were scattered across the southern latitudes. He quickly came to realize that landmasses which were close to each other did not always have similar plants: for example, the gum trees so common in Australia were not found in New Zealand. By contrast, islands that were separated by vast stretches of ocean sometimes had similar species. There were irregularities in these patterns. The plants of Kerguelen Island, for example, appeared to be closely related to those of distant Tierra del Fuego, not to those of the Auckland Islands, which were geographically closer.[4]

*Joseph Dalton Hooker.*

Upon his return to England, Hooker would continue to build his collection of southern flora with the help of botanizing friends in New Zealand and Australia, and also to develop his ideas about plant distribution. Between 1844 and 1859 he published, in three volumes, his magnum opus, the snappily titled *The Botany of the Antarctic Voyage of H.M. Discovery Ships Erebus and Terror in the years 1839–1843, under the Command of Captain Sir James Clark Ross*. In Hooker's mind, the pattern of plant distribution across the southern landmasses indicated they had once been connected, possibly by land bridges of some kind. He envisaged a 'continuous extensive flora … that once spread over a larger and more continuous tract of land'.[5] This concept — that species have arisen from populations that were once joined but have now been separated by a geographical barrier, such as a mountain range or an ocean — is known as vicariance.

In letters, Hooker communicated his ideas to his mentor and friend, Charles Darwin. Along the *Beagle*'s voyage, Darwin had also been fascinated by the presence of closely related species on disparate landmasses. *Nothofagus*, or southern beech, for example, is a genus Darwin observed proliferating in South America, Australia and New Zealand, despite the huge stretches of ocean that separate these places. Darwin agreed with Hooker that understanding the distribution of plants and animals around the world would ultimately be 'the key which will unlock the mystery of species'.[6] However, he strongly disagreed with Hooker's idea of vanished land bridges. 'It shocks my philosophy to create land, without some other independent evidence,' he wrote to Hooker.[7]

Darwin was a firm proponent of what is today called dispersal, the ability of plants and land-based animals to cross vast stretches of oceans by floating or flying (either by being borne on winds or by hitching a lift with birds). On his voyage aboard the *Beagle*, Darwin had noted how trees of the genus *Sophora*, an attractive tree with bell-like yellow flowers, were found all across the southern latitudes. Later, back in England, he conducted tests to try to prove the ability of *Sophora* and other seeds to survive immersion in salt water. (Darwin's children cheered each successful post-immersion germination as another victory over Hooker.)[8] In Darwin's mind, and from his observations in places like the Galápagos Islands, oceans should be no barrier to species crossing them — especially in the vastness of geological time, the scale of which was by now becoming apparent to him. This friction, between Darwin's dispersal and Hooker's vicariance, would smoulder away in the background of science for the next two centuries. It would have a direct bearing on the search for Gondwana.

Darwin, meanwhile, was on the brink of his epoch-making insight into the way in which species evolved through time. In 1858, along with fellow naturalist Alfred Wallace, who had independently arrived at the same insight, he presented his theory of evolution by natural selection to the Linnaean Society in London.[9] The following year, he published his book *On The Origin of Species*, which laid out these monumental insights in full. Its publication would shake science to its very core, providing the world with a glimpse of the enormity of geological time and the incredible mystery of life's journey on this planet.

The door to deep time had been cracked open by Buffon, Cuvier, Hutton, Lyell and many more who preceded them. Darwin and Wallace gave it an almighty shove. And it was here, in deep time, that we would at last find our lost southern land.

# EDUARD SUESS' MASTERWORK

If you're going to write a book entitled *The Face of the Earth* it might as well be a tome. As it was, *Das Antlitz der Erde*, Eduard Suess' masterwork, clocked in at four volumes and over 2000 pages. Published in stages between 1883 and 1909, it took him 26 years to write and synthesized a lifetime of hard-won theories on the structure of the Earth's crust and the placement of its oceans.

The size of the book reflected the enormity of his topic. 'We are prone to forget,' Suess wrote, 'that the planet may be measured by man, but not according to man.'[10] Suess explored the European Alps in thorough detail, analyzing their folded geology. The accepted wisdom at the time was that mountains were raised by vertical geological processes, something Suess profoundly disagreed with. In

*Eduard Suess.*

essence, *The Face of the Earth* is an enormous targeted missive whose sole purpose is to tear down this orthodoxy.[11] Well acquainted with the geology and fossils of his native Alps, Suess realized that these mountains had been thrust up not by vertical movements in the Earth's crust but by lateral ones. The land, he saw, had crumpled like a blanket before enormous pressures — but what caused these pressures?

Like many of his time, Suess believed the Earth was cooling, and shrinking as it did. As it contracted, he thought, the continents were thrust up against and upon one another.

In his book, Suess explored geological and biological connections between the continents, paying close attention to the startling connection between the geology of South Africa's Karoo and the Gondwana region of India. Speaking of the glacially deposited sediments of the Talcher Formation, which underlie the Gondwana sequence, he wrote, 'The resemblance to the Dwyka conglomerate, which forms the lowest division of the African Karoo series, is very striking.'

Suess then goes on to describe the three units of sediment that overlie the Talcher:

*An undeniable resemblance exists between the structure of South Africa and that of the Indian peninsula. Each of these two great regions has remained for a long time — Africa since the Carboniferous at least, India probably since the same period undisturbed by any manifestation of tangential force; there has been no folding of the mountains, each is veritable table-land.*

*In both cases we find, resting upon an older foundation, a mighty series of non-marine deposits which extend from the Permian to the Rhaetic and perhaps into the Lias [Jurassic age].*

Suess also noted the arresting fact that *Glossopteris* fossils had been found in both these places, as well as in Australia and South America. In all these places, *Glossopteris* leaves were typically found as the dominant species in a diverse assemblage of plant fossils that included ginkgos, conifers, ferns, horsetails, sphenophytes, ferns and mosses. Lycopods — relatives of modern-day club mosses — were also common in these fossilized forests. This collection of plants represented an ancient ecosystem, a vanished forest that seemed to have sprawled across a number of widely separated countries. Collectively, this fossilized forest became referred to as the *Glossopteris* flora.

Suess saw the presence of the *Glossopteris* flora in these disparate continents as evidence for the existence of a former supercontinent. 'We call this mass Gondwana-Land,' he wrote, 'after the ancient Gondwana flora which is common to all its parts.' Like Hooker, Suess believed land bridges had connected these pieces of the supercontinent, allowing animals and plants to cross between them. Suess believed the shrinking of the Earth had caused parts of Gondwanaland to sink beneath the waves.

> *Throughout this long interval, a series of similar terrestrial floras, accompanied by peculiar land reptiles, flourished in both regions. Then came collapse. A new ocean was created and the continents assumed other forms.*
>
> *It is not only in the great mountain ranges that the traces of this process are found. Great segments of the earth's crust have sunk hundreds, in some cases, even thousands, of feet deep, and not the slightest inequality of the surface remains to indicate the fracture ...*
>
> *Time has levelled all.*[12]

While his ideas of vanished land bridges would in time be disproven, Suess' extraordinary insights into the Earth's geological processes were so fundamental to the development of the science that his achievements have become perversely invisible — part of the accepted background knowledge. As one of his biographers, Ali Mehmet Celâl Şengör, puts it, 'His work has become so much a part of what geologists do and encounter every day that citing his name has long been considered superfluous.'

And yet, when Suess published *Das Antlitz der Erde* there was still an awfully long way to go before the significance of his insights would become widely accepted. Mainstream geology was about to enter a long, dark age of denial.

# THE AGE OF THE EARTH

How old is the ground we walk upon? This was the fundamental question that loomed over nineteenth-century geology.

It was a gathering sense of the extreme antiquity of the Earth that led geologists of the nineteenth century, most prominently James Hutton, to propose an infinite age for the planet.

Darwin, in his *On the Origin of Species*, had proposed a unit of marine rock strata in southern England known as the Weald had taken 300 million years to erode to its current level, a figure that suggested an age for the Earth that was much, much older than that again.[13]

Among those who disagreed with Darwin was the famed mathematician William Thomson, better known as Lord Kelvin. The Glaswegian, who was regarded as an expert on thermodynamics, felt Darwin's estimated age for the chalk was 'absurd'.[14] Based on his calculations on the thermal conductivity of rocks, Kelvin calculated the age of the Earth to be around 20 million years. His figures were refuted by most geologists, but Kelvin was one of the most highly regarded scientists of his day, so few dared to challenge him publicly.

However, experimental evidence from another quarter would soon trump Kelvin's claims. It came with the discovery of radioactivity by Henri Becquerel in 1896 and its subsequent study by Marie and Pierre Curie.[15] Uranium is a rare element in the Earth's crust, a remnant of the rain of fire that formed the Earth in its molten infancy. At its core, uranium has a nucleus containing a whopping 92 protons. This densely packed nucleus is unstable — and as a result, uranium radiates atomic particles at a steady rate.

The great physicist Ernest Rutherford discovered that radioactive elements like uranium break down over time to form new elements in a predictable manner. Inspired by this understanding, an American scientist called Bertram Boltwood realized that by knowing the half-life of uranium — that is, the amount of time it takes for half the amount of the uranium in a rock sample to decay to lead — and then by measuring the relative amount of uranium and lead in the sample, it would be possible to figure out how old the rocks were. In other words, you could use uranium to estimate the age of the Earth. Boltwood examined uranium-bearing samples from some of the oldest rocks on the planet and came to the conclusion that Earth was an astonishing 2.2 billion years old.[16] As we now know, Boltwood's estimate was still short by a good couple of billion years (modern estimates put the age of the Earth at around 4.5 billion years)[17] but this was a definite improvement on Lord Kelvin's paltry 20 million.

At last, geologists had the amount of time they needed for their grand processes to unfold. The stage was set for Suess' Gondwanaland to be fully imagined. But to really connect the dots of the Austrian's theory would require evidence from the furthest reaches of the planet — a place only the most intrepid dared to tread.

## FOSSILS IN ANTARCTICA

Carl Anton Larsen was born to hunt whales. The Norwegian mariner first went to sea at the age of nine, hunting seals with his father, and by his early twenties had acquired his first whaling ship, which despite being almost wrecked on her first voyage would soon be filled to the gunnels with the flesh of bottlenose whales hunted off the coast of Norway.

By the end of the 1800s the world's stocks of whales, at least those considered commercially viable to hunt, was running low. Right whales, the most easily hunted whale, had been brutally harvested throughout the Southern Hemisphere and were now facing extinction. The fast-moving rorqual whales — the humpback, fin and blue whales — were still too elusive to hunt using the technology of the day. The whalers of Norway needed more right whales to hunt, so they began looking further afield. Cook and Ross had returned from the Antarctic with reports of huge amounts of whales in those waters, so it was to Antarctica that Norwegian whaling companies turned their attention.[18]

In 1894, the 30-year-old Larsen captained the whaling ship *Jason* on a trip to the barely explored Antarctic Peninsula. The expedition entered the Weddell Sea and penetrated as far south as the ice shelf that would one day be named the Larsen Ice Shelf, after its captain. There, the adventurers found themselves surrounded by countless whales — just not the ones they were seeking. 'Blue whales frolicked in countless shoals,' wrote Larsen, 'as well as the humpback, but there was no sign of the characteristic blowing of the right whale.'

The expedition stopped at Seymour Island, an ice-free island off the Antarctic Peninsula. With no right whales to hunt, Larsen made the decision to fill the ship's hold with seal blubber instead. When they went ashore on Seymour in search of seals to hunt, Larsen and his men made an astonishing discovery. Eroding out of the sedimentary rocks, as if they had been buried there in the last king tide, were chunks of fossilized wood. This was a momentous find: these fossils, the first ever recovered from Antarctica, revealed for the first time that trees had once grown there — the continent had not always been a frozen wasteland.

Later in the voyage, the *Jason* and her crew met up with another whaling expedition, the Dundee Antarctic expedition. This expedition had two scientists on board, who had been frustrated by the whalers' commercial interests continually overriding their scientific aims.

*The* Antarctic *in pack ice.*

They'd had little opportunity to look for fossils. Seizing the chance to get their hands on items of such immense scientific interest, these scientists traded tobacco for most of the fossils aboard the *Jason*. Larsen, however, having a strong personal interest in science, held onto his finds, taking them home with him.[19] He returned to Norway with a hold full of valuable seal oil and skins and a haul of new scientific knowledge of the Antarctic Peninsula.

As a whaling venture, the *Jason*'s expedition might have been a failure, but Larsen's plant fossils fuelled interest in launching a true scientific venture to the region. The Swedish explorer Otto Nordenskjöld asked Larsen to captain another expedition, aboard a ship called *Antarctic*. Also on board, along with a number of other scientists, would be the Swedish paleontologist Johan Gunnar Andersson.[20] In 1901, the *Antarctic* called in at the South Shetland Islands, on the west side of the Antarctic Peninsula. Finding no fossils there, they moved around the east side of the Peninsula and set up their base for the winter at Snow Hill Island, where ammonite fossils had been found. The ship left an overwintering party, led by Nordenskjöld, on Snow Hill Island and returned to the Falkland Islands to wait out the winter.

On the Falklands, Andersson went looking for fossils. At Port Louis, he collected brachiopods from the same Devonian sandstone in which Darwin had found them seven decades earlier. He also collected plant fossils — among them *Phyllotheca*, an ancient Permian horsetail.[21] He was not blind to its significance. *Phyllotheca* is a key member of the *Glossopteris* flora. It had been found in India, South Africa and Australia, alongside the glossopterids and their Permian allies. Did this mean that the Falklands, too, was a member of Suess' great southern land? It was a question that would linger long after the Swedish expedition had reached its dramatic climax.

When spring returned, the *Antarctic* returned to the Antarctic Peninsula to collect the overwintering party. Larsen, however, was unable to guide his ship through the thick pack ice to reach Snow Hill Island. Andersson and two others were put ashore at nearby Hope Bay, from where they planned to sledge across the sea ice to Snow Hill Island to retrieve the shore party.

However, by now the sea ice had broken up, so they were unable to make it to Snow Hill Island, and so returned to Hope Bay. While they were making this attempt, the *Antarctic* had become trapped in the drifting ice, and after floating helplessly for a month it was crushed, and sank.

Larsen and his crew took to the lifeboats and made for nearby Paulet Island, where they erected a meagre stone hut. They spent the entire winter there, surviving on penguin and seal meat. The other two groups at Snow Hill Island and Hope Bay likewise hunkered down for a brutal winter.

Incredibly, all three groups made it through the darkness of the Antarctic's long night to be reunited the following spring.

The Paulet Island group, led by Larsen, made it to Snow Hill Island just days before the rescue expedition sent from Argentina was due to give up on finding them and return home.

Larsen and his crew were taken back to Buenos Aires. Having survived such an ordeal, Larsen might have been forgiven for never wanting to see Antarctica again. Instead, he immediately set about raising finance for a venture to return to the Peninsula to hunt the whales he had witnessed in their thousands there. Larsen would go on to pioneer the whaling industry in Antarctic waters, but his name is etched in history as a fearless Antarctic explorer and the man who first discovered the first plant fossils in the frozen land.

Built as a whaler in the shipyards of Dundee, *Terra Nova*'s sturdy construction made her ideal for the Antarctic voyage Robert Falcon Scott had in mind.

Scott had already led one famous expedition to the ice between 1901 and 1904, aboard the ship *Discovery*. That expedition had made a valiant attempt on the South Pole, but the team had been forced to turn back after the dogs they were using to haul their sledges proved to be less efficient than they had hoped.[22]

Upon his return to Britain, Scott began raising interest in a return trip. The pressure to be the first to get the pole was intense and Scott was well aware of other expeditions being planned by Japan and

Norway. He travelled around the United Kingdom meeting potential sponsors and trying to raise funds for the voyage. Among those he met on these travels was Marie Stopes, a well-known author of feminist books, including the highly controversial sex manual *Married Love*. Stopes was also a renowned paleobotanist with a particular interest in Carboniferous plant fossils. After his initial meeting with Stopes, Scott went to visit her at the University of Manchester, where she was a professor. At this visit, it's likely Stopes impressed on Scott the importance of finding *Glossopteris* in Antarctica. *Glossopteris* would be the missing piece in Suess' jigsaw — the last piece of evidence that would finally connect the landmasses of Gondwanaland. Stopes apparently asked Scott if she could join his expedition. This did not come to pass. However, Scott's interest in finding fossil plants was certainly piqued.[23]

From day one, the *Terra Nova* expedition had been planned not only as a mission to claim the glory of being first to reach the pole for Britain, but also as a quest of scientific discovery. Geological and meteorological studies were planned, and exploration of uncharted parts of the continent was also a major objective. 'No one can say that it will have only been a pole-hunt,' expedition scientific chief Edward Wilson wrote. 'We want the scientific work to make the bagging of the pole merely an item in the results.'[24]

With the money finally raised, the expedition departed Britain in 1910 and made its way to New Zealand, the launching point for the Antarctic voyage. The ship loaded up in Port Chalmers and headed out onto the Southern Ocean, with a hold full of supplies and animals including dogs and ponies.[25]

*Terra Nova* struck a big storm on the way south, which resulted in the loss of some ponies and supplies. Upon arrival in the pack

*Antarctic plant fossils recovered by Robert Falcon Scott and his team, Natural History Museum, London.*

ice, the ship was held fast for 20 days, significantly holding up the exploration schedule.

Finally, the expedition set up base at Cape Evans and began exploring King Edward VII Land to the east. It was on one of these explorations that it was discovered that they were not alone: a rival Norwegian expedition, led by Roald Amundsen, was also camped on the ice edge, preparing for their own assault on the pole.

What was left of the summer was spent laying out supply depots for the polar journey the following season. Pushing to reach the final depot in the teeth of a blizzard and against the wishes of fellow team member Lawrence Oates, Scott elected to lay the final depot almost 50 kilometres (31 miles) short of where it had been intended — a decision that would have crucial, fatal consequences.

In October 1911, the pole party departed Cape Evans. Ponies and dogs were used to haul the equipment to the base of the Beardmore

Glacier, at which point the ponies were shot for meat and most of the party returned to Cape Evans with the dogs.

Scott and four others pushed on to the pole. After 20 days of hard travel, they reached their goal, only to discover Amundsen had got there first. They turned and headed for home, visions of England dancing before them as they trudged through the sastrugi. After stopping to collect 16 kilograms (2½ stone) of plant fossils at the top of the Beardmore Glacier, they hauled their sledge down the glacier, and into oblivion.

A year later, on 10 February 1913, the *Terra Nova* arrived off the coast of New Zealand.

Two sailors launched a small boat and rowed into the port of Oamaru. There, they cabled England with the awful news that they had returned to civilization without their leader and the rest of the polar party. The next day, news of Scott's demise broke in the newspapers of Britain and the world.

The *Terra Nova* returned to England, where the fossils Scott and his team had collected on the Beardmore Glacier were sent to Cambridge University. They were examined by the great botanist Albert Seward, who formally described some as *Glossopteris indica.* At last, as had been predicted, the iconic flora of Gondwana had been recovered from the most unlikely place on Earth imaginable.

In a way, it was a fitting vindication of the *Terra Nova* expedition. Scott's quest to be first to the pole had come to naught, and yet the fossils he and his men collected on the Beardmore Glacier that day had helped solve one of the great mysteries in science.

The final piece of Suess' puzzle was now in hand. 'Gondwanaland' was looming into view. And yet it would take another long half-century before science would finally come around to accepting it.

# JOINING THE JIGSAW PIECES

*Alfred Wegener.*

At the same time as the doomed Scott was trundling across the Antarctic ice, another polar explorer was preparing for his own voyage. The German meteorologist Alfred Wegener was headed to Greenland on an expedition of science and exploration.[26] Wegener, a high-achieving polymath with a penchant for Arctic travel and hot air ballooning, was not your average academic. In fact, his broad knowledge and wide-ranging interests made it hard for him to find an academic role.

The previous year, as an (unsalaried) tutor at the University of Marburg in Germany, Wegener had stumbled on an article that listed fossil plants and animals found on both sides of the Atlantic, and suggested land bridges once connected the two continents. This phenomenon intrigued Wegener.[27]

For hundreds of years, ever since a cohesive map of the world had emerged, it had not escaped people's attention that some parts of the world's continents appeared to fit together as if they had once been connected. Wegener decided the continents looked as if they had once fitted together because they had.

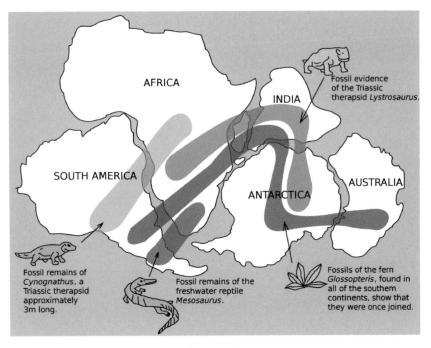

Fossil evidence of the Triassic therapsid *Lystrosaurus*.

AFRICA

INDIA

SOUTH AMERICA

AUSTRALIA

ANTARCTICA

Fossil remains of *Cynognathus*, a Triassic therapsid approximately 3m long.

Fossil remains of the freshwater reptile *Mesosaurus*.

Fossils of the fern *Glossopteris*, found in all of the southern continents, show that they were once joined.

*Wegener's vision of Gondwana, showing linkages between continents.*

He was by no means the first to suggest this. Abraham Ortelius, a Dutch cartographer, had noted as much in 1596, theorizing that the Americas had been torn away from Europe and Africa '... by earthquakes and floods'. The famed explorer, mapper and natural scientist Alexander von Humboldt had also noted the similarity between the coastlines of the two continents in the early 1800s and suggested a former connection between them, as did the Italian geographer Antonio Snider-Pellegrini, who in 1858 proposed all the continents had been joined in Carboniferous times, only to be separated by the biblical flood.

But whereas these previous musings had been formulated with little scientific evidence, Wegener threw every ounce of his considerable scholarly heft behind his ideas. He proposed all the world's continents had once been joined in a conglomeration he called Pangaea, which means in Greek 'all Earth'. He painstakingly assembled evidence for this former connection, comparing plant and animal species that lived on either side of the Atlantic and Pacific, and pointing out how similar the fauna of Australia and South America, for example, was.

He also gathered evidence to show how rock strata across the world joined up like pieces of a jigsaw puzzle — the Appalachian mountain range on the east coast of North America appeared to have once been connected to the mountains in Ireland, Britain, Greenland and Norway. Likewise, mountains in Brazil and Ghana also appeared to be an exact geological match.

Then there were ancient fossils like *Mesosaurus*, an early freshwater crocodile-like reptile whose fossils are found in South America and Africa. Because *Mesosaurus* lived only in freshwater, Wegener asserted, there was no way it could have swum the Atlantic.

And then, of course, there was the *Glossopteris* flora, whose appearance across Australia, South America, Africa, India and now Antarctica spoke to Wegener of an indisputable former connection.

Wegener pointed out that these deposits were evidence of a climatic region that had once existed in a belt around certain southern latitudes. 'We are dealing here with bogs of the southern subpolar rain belt,' he wrote, 'formed in just the same way as the ... peat bogs of northern Europe and North America.' He continued:

> *These regions must have been joined because they now include far too large an area for their climate of that time to have covered.*
>
> *The distribution of the* Glossopteris *flora forces one to assume a former large area of dry land joining the southern continents.*[28]

Like Suess before him, Wegener differentiated between the two different kinds of crust: sial, a lighter crust of which the continents were composed, and sima, the denser crust of ocean floors. He proposed that continents made of sial ploughed through oceanic crust like ships through the ocean.

In 1912, Wegener presented his ideas of 'continental displacement' to the Frankfurt Geological Association, to muted response. Three years later, as he recovered from wounds sustained while fighting for Germany in World War I, he synthesized his ideas in a book called *The Origins of Continents and Oceans*, which was published in 1922. Perhaps fuelled by anti-German sentiment in the aftermath of the war, scientists from around the world fell on Wegener's provocative theories like wolves upon a wounded calf, lambasting them as 'German pseudoscience'. The American geologist Rollin T. Chamberlin led the savaging in the United States, accusing Wegener of taking 'considerable liberties with our globe'. Even German scientists turned on their countryman, attacking his 'delirious ravings'.

The reason Wegener's ideas received such a dismissive reaction are complex. For one thing, Wegener was unable to provide a convincing mechanism for how continents might be able to move across the face of the Earth. His ideas of continents moving through oceanic crust were also shown to be implausible. Other aspects of Wegener's work were also off the mark, such as his suggestion that the break-up of Pangaea was driven by centrifugal forces generated by the Earth's spin.

But perhaps the biggest reason it took so long for continental drift to be accepted was unwittingly encapsulated in a criticism levelled at Alfred Wegener by Chamberlin in 1926. 'If we are to believe Wegener's hypothesis,' Chamberlin is famously quoted as saying, 'we must forget everything which has been learned in the last 70 years, and start all over again.'[29]

Wegener responded to the assault by refining his theories and republishing them. He was convinced of the rightness of his ideas and knew it was only a matter of time before he would be vindicated. In 1930, at the age of 50, Wegener embarked on another Greenland expedition, this time to rescue two stranded researchers. He ferried lifesaving supplies to the stricken team and then, realizing there were not enough supplies to sustain all of them, headed out on the return journey to base camp.

Like Captain Scott before him, Wegener never made it, succumbing to the cold of another polar realm. He would never live to see his most famous idea validated.

# 550 MILLION YEARS AGO

The super-plume beneath Rodinia grows in intensity, fuelled by heat trapped under the continent.

Rodinia fractures along pressure gradients, the broken chunks sliding off the areas of high pressure in the centre and moving away towards areas of lower pressure at the edges.

These continental fragments of Rodinia now head off across the surface of the planet on their own journey. Along their leading edges the thinner, denser, oceanic crust they encounter is subsumed. It slides down before the bulk of the onrushing continent, diving hundreds of kilometres into the interior of the Earth.[30]

Down in the mantle, this old seabed has become part of the lubrication zone, the plastic sleeve of the Earth's interior. This is a realm where human understanding of time falters — where movements and processes take on a monolithic slowness. It's a realm of deep Earth rock; peridotite and eclogite, rock which moves like liquid through the immensity of geological time. Like any liquid, this mantle rock has its fluctuations and moods, currents that writhe inside it over millions of years. It is these that drive the ceaseless movement at the surface.

The trailing edges of the diverging continents draw new oceanic crust with them, hauling it like a tablecloth from long cracks in the Earth. This new crust is hot and buoyant, and forms massive ridges along the sea floor where it extrudes. These ridges bulk out the sea floor, causing sea levels to rise.

As the newly born continents drift apart they open up new, shallow seas, providing niches and habitats for evolution to occur. And so life begins to radiate in fantastic ways. Weird soft-bodied creatures, reminiscent of sea fans and worms, now inhabit the planet's oceans. The land, however, remains bare — life has yet to find a way to thrive on its harsh, rocky surface.

Rodinia is now a distant memory. She has fragmented into Laurentia, Baltica, Siberia, and the blocks of crust that will one day form Gondwana. Her passing is part of the pulsing rhythm of the geology of Earth, for supercontinents contain within them both the mechanism for their own creation and the seeds of their own demise. The Earth keeps a record of their existence in remnants of twisted, gnarled rock.

The truth these fragments reveal is this: Rodinia was not the first. She will not be the last.

*The Trans-Gondwana Mountains.*

And then, the cycle reverses. The carpets of crust underlying the oceans that opened up with the break-up of Rodinia age and cool, and as they do they become denser and sink.[31]

At the edges of the continental fragments that now lie scattered across the surface of the Earth, this oceanic crust begins to be subsumed, its sheer mass drawing it down like a gumboot full of water. It slides under the continental crust at a steep angle, pouring into the mantle below. In doing so, it draws the continents together, as the oceans between them vanish in the crush.

When finally all the oceanic crust is subsumed, continents are drawn together once again. Where continents meet, subduction ceases. It is impossible for one to slide under the other. Instead, they pile into one another with unimaginable, slow force. In the Southern Hemisphere, two huge chunks of continent — slabs that will one day be known as East and West Gondwana — meet in what may be the most intense collision that has occurred on Earth in the last billion years.

The collision completes the southern supercontinent Gondwana. The mountain range it thrusts up is 8000 kilometres (5000 miles) long and 1000 kilometres (621 miles) wide — a range that dwarfs anything on Earth today. It spans the width of the newly formed supercontinent.[32] The endless forces at work beneath the Earth push the mountains high into the atmosphere, where the wind and rain begin their slow, irrepressible grind.

Weather systems, fat with ocean evaporation, sweep inland off the Panthalassic Ocean, where they are blown up against the bulk of the Trans-Gondwana Mountains in the deep interior. Forced upward by the contours of the land, the winds relinquish their consignment of water in the thinning air. Rain pours from the sky, scouring away at this vast mountain range, eroding and stripping their slopes. Naked rock is left exposed to the elements, allowing chemical reactions to take place. Running water strips minerals from the rock — calcium, phosphorus and iron. With no plant life to hold this material in place, the scale of erosion is immense. Unconstrained, sheets of water that are not so much rivers as terrestrial tsunamis course off her sides, carrying all of this material to the coast.

The supercontinent's huge bulk pushes up against the sky, and down on the Earth's crust. Because the oceanic crust that surrounds it is old and dense, it has sunk and sea levels are lowered. The wide continental shelves around Gondwana have become shallow, with light from the sun penetrating all the way to the sea floor, creating a benign habitat for life. The huge amount of minerals pumped out by the steadily eroding continent fuel productivity in these warm, rich seas. As a result, marine creatures are evolving at an unprecedented pace. From simple, unicellular, colonial animals and soft-bodied sessile creatures, a horde of new species, genera, families and phyla evolve in a geological instant.

In just a few million years, most modern forms of life appear: creatures with spinal cords, compound eyes and jointed legs. Armed with eyes and jaws, the first predators evolve and, in response, those animals that are the hunted must develop defenses — hard shells manufactured from calcium which, thanks to the erosion of those great mountains, the ocean is now rich in. An arms race is underway.

These are the first complex animals on Earth, creatures with hard body structures that fossilize as they die, creating a detailed record of this 'Cambrian explosion'.

# 4

# ACCRETION

The Pacific Ocean fusses at the shore of my country, gently eating us one mouthful of sand at a time. The ocean brings us our rain, our winter storms and summer breezes. Our ancestors crossed this ocean — the Polynesian forerunners of Māori traversing it aboard double-hulled waka; mine on sailing ships from the far side of the world.

The Pacific is the biggest ocean on the planet, spanning almost 20,000 kilometres (12,427 miles) at its widest point. And yet it was once much, much bigger.

Only in the last century have we at last begun to peer into the Pacific's watery eye. And when we did, we at last found the truth about our restless mother, Earth.

## 'THE SUPPORTERS OF CONTINENTAL DRIFT'

Despite the howls of indignant dissent and ridicule, Wegener's theories of continental drift found their supporters too. Among the most prominent was the South African geologist Alexander du Toit, who recognized the battle lines over continental drift had well and truly been drawn. 'The principles advocated by the supporters of continental drift form generally the antithesis of those currently held,' he wrote. 'The differences between the two doctrines are indeed fundamental and the acceptance of the one must largely exclude the other.'[1]

In South Africa, Du Toit had mapped the Karoo system and was intimately familiar with its distinctive succession of rocks. In 1923, he was given a grant to compare the geology of South America and South Africa. The report he produced, based on studies in Argentina, Paraguay, Uruguay and Bolivia, showed that plant and animal fossils in these countries were the same as had been found in the Karoo rocks, and not only that, they appeared in exactly the same sequence.

In 1937, Du Toit published a book called *Our Wandering Continents*, which he dedicated to Wegener. In the book, he hypothesized that Pangaea, the landmass that Wegener had proposed, had separated into two pieces around 205 million years ago — Gondwanaland in the south and Laurasia in the north. Gondwanaland, he proposed, had subsequently separated into South America, Africa, India, Australia and Antarctica.[2]

Meanwhile another Wegener convert, British geologist Arthur Holmes, tackled the great weakness in Wegener's theories: the lack of a plausible mechanism by which continents could move in relation to one another. Like Bertram Boltwood, who had previously estimated the age of the Earth using radioactive decay, Holmes was interested in how elements break down through time. He understood that after its fiery beginning, the centre of the Earth was losing heat as elements in the interior were broken down. Holmes suggested that heat loss in the Earth's core would cause convection currents in the mantle. In the same way that tea leaves circulate through a freshly brewed cup, molten rock would rise buoyantly through the mantle, then when these currents struck the bottom of the Earth's crust they would move sideways. As they subsequently cooled, they would sink again.[3] This constant movement, Holmes thought, was shunting the tectonic plates around, causing the continents to move with them.

For Wegener and Du Toit, the presence of *Glossopteris* on the Gondwanan landmasses was strong evidence for their former connection. But not everyone agreed. The Yale geologist Chester Longwell, for example, took exception to the use of *Glossopteris* as evidence for continental drift. 'Consideration of some problems presented by distribution of modern plants should give us pause ... in accepting the strong claims based on ancient floras, including the famous *Glossopteris* flora,' he wrote.

Longwell went on to point out that some modern plants may have been able to traverse the Pacific by floating or hitching a lift with migrating birds. *Glossopteris*, he proposed, might well have been capable of crossing oceans to colonize landmasses, and therefore was not reliable evidence for their former connection. 'Emphasis that has been placed on the *Glossopteris* flora as "compelling evidence" of once-continuous lands,' he wrote, 'seems dangerously near the unscientific procedure of selecting evidence to support a favoured theory.'[4]

While Wegener's theories were discarded by most teaching institutions in Europe and the United States, there was one country in which they were taught — South Africa. Among the generation who grew up being exposed to the ideas of Wegener and Du Toit was the paleobotanist Edna Plumstead.

Plumstead studied *Glossopteris* fossils from the Karoo, and in 1956 discovered the first *Glossopteris* sexual organs. Plumstead's paleobotanical research in Antarctica and elsewhere convinced her that the presence of the *Glossopteris* flora across the globe was incontrovertible evidence for continental drift.

While acknowledging the possibility that *Glossopteris* might have been capable of dispersing across oceans, she pointed out that the current distribution of its fossils occurred across too great a latitudinal span to have grown at those latitudes. She wrote:

> *How can we explain the fact that the fossil plants of peninsular India are closely comparable, and often identical with those of Australia, Africa south of the Sahara, Argentina and Brazil but above all, with those found in the heart of Antarctica at 86°S where today no vascular plant could live?*
>
> *The presence of whole and almost identical floral assemblages in such widely separated areas defies explanation by any means other than that the continents themselves have moved and were once close together.*[5]

The great Indian geologist Birbal Sahni defended continental drift throughout the 1930s, based on his observations of fossilized forests in both India and neighbouring China. The *Glossopteris* forests found in northern India, he realized, contrasted with the *Gigantopteris*-dominated flora found in fossil deposits in China. The two floras had clearly grown in two very different climate regimes, and to Sahni this was evidence that India had subsequently been moved into its present position abutting China by geological forces.[6]

Despite the compelling arguments of Wegener, Du Toit, Holmes, Sahni and Plumstead, the idea of continental drift was very much relegated to the margins of scientific enquiry, scorned as a fringe view with no scientific merit.

Between 1925 and 1965, geology was stuck in what Ali Mehmet Celâl Şengör describes as the 'Dark Intermezzo', a period when the scientific establishment refused, or was unable, to grasp the importance of these ideas.[7] According to Şengör, it was partly a problem of perspective. Whereas scientists of Suess' time had been generalists, scientists in the early part of the 20th century had become increasingly specialized. Many geologists were focused on narrow areas of study, which made it difficult for them to step back and see the bigger picture of plate tectonics. 'Geologists lost the forest for the trees,' he wrote. 'Because they have come to know more and more about less and less, they have become intolerant of criticism, coming from outside their narrow sub-fields. This created a vicious cycle of ever narrowing specialties and ever increasing dogmatism ... geologists lost sight of the planet.'[8]

It was only when new technology allowed us to look deep into the ocean that the mainstream scientific establishment was at last forced to take continental drift seriously.

In 1925, Germany was in economic ruins. The cost of World War I and the crippling sanctions imposed upon the country by the Treaty of Versailles had left it destitute. Inflation was skyrocketing and the country was going broke.

That was when German chemist Fritz Haber came up with a radical idea to pull his country out of the financial mire. He had read that sea water contained a large amount of gold in its chemistry and proposed that Germany could solve its problems by extracting it.[9] The German government bought the idea and financed a ship to venture into international waters to give it a shot.

The ship was equipped with the latest echo-sounding technology, capable of mapping the deep-sea topography. In the course of its journeys, the ship's crew made an interesting discovery. The Atlantic Ocean, it transpired, was split down the middle by a long, wide ridge. The significance of this finding to the raging storm that surrounded continental drift would take another couple of decades, and another war, to become apparent.

The gold in sea water would remain elusive.

Dawn broke on 19 February 1945 to reveal a fearsome flotilla of US military ships standing to, just off Iwo Jima's seething beach.

Over the next four weeks, troop carriers would pour thousands of soldiers onto the Pacific island, right into the teeth of Japanese guns that pounded the beach. By the time it was all over, more than 6000 Americans and 22,000 Japanese would have died, and the Americans would have won a major turning point in World War II.

At the helm of the American attack transport ship USS *Cape Johnson*, which would provide support to the troops throughout the campaign, was Harry Hess who, outside of wartime, was a professor of geology at Princeton University. With the Iwo Jima bloodbath over, the *Cape Johnson* headed off across the Pacific towards its next engagement. It was equipped with relatively new sonar technology, which allowed the US Navy to hunt submarines. Sonar also had the unintended consequence of allowing geologists to accurately map the sea floor. Fascinated by what sonar revealed of the unexplored sea floor, Hess would leave it running as the ship travelled across the ocean. During the ship's many journeys, Hess built up a detailed picture of some of the geological features of the deep seabed.

Instead of a flat, featureless bottom, the ocean, it transpired, was riven with trenches and pocked with undersea mountains. It became clear that there was active geology going on under the sea. Mid-ocean ridges ran around the entire planet like the seam on a tennis ball, and cut down the middle of all the world's oceans. These ridges were seamed with a gulf or canyon that ran down the centre. The further from the ridges you went, the deeper the sea became.

These observations led Hess to believe that the sea floor was being formed as liquid magma at these ridges, then moving away from the ridges on either side, spreading like a conveyor belt in opposing directions and sinking as it did. When it struck the edge of a continent this dense, cold oceanic crust was being subsumed back into the Earth. This would explain why fossils found on the sea floor were never more than 180 million years old — anything older had been swallowed up at these subduction zones.

'Sea floor spreading' offered an explanation for how continents could move across the planet — a plausible mechanism for Wegener's continental drift. It wasn't the continents that were drifting or ploughing through oceanic crust as Wegener had proposed; it was the sea floor that was moving, carrying the continents apart.[10]

In the 1960s the theory was strengthened by dating of ocean-floor core samples, which confirmed the ocean crust was youngest at the ridges and became progressively older the further on either side from it you went.

Indisputable evidence for sea floor spreading was finally secured in 1963 by an ingenious observation based on magnetic anomalies.

Throughout the planet's long history, Earth's geomagnetic field has periodically flipped, with north and south switching places. Geomagnetic reversals are relatively common — there have been eighteen in the last 80 million years alone. These reversals appear to have minimal impact on life on Earth. In fact, the only way we know about them is that evidence for them is recorded in rocks. When igneous rocks are first formed, iron-bearing minerals in the molten lava act as tiny magnets, aligning to the direction of polar north in the instant before the rock solidifies. Magnetic studies in the 1960s showed that oceanic crust was composed of parallel bands of alternating magnetic polarity, which when depicted in a diagram appear as stripes on the sea floor.

This information was independently seized upon by two scientists, the American Frederick John Vine and Canadian Lawrence Morley. Vine and Morley realized that if Hess' theory of sea floor spreading was correct, evidence for it should be preserved in just such a way in the magnetic record. As newly formed basalt solidified at the mid-ocean ridge, the magnetic alignment of the geomagnetic field at that time would be preserved in the iron minerals like a tape recorder. Over tens of millions of years, as the sea floor spread away from the ridge, evidence for these reversals should be preserved as alternating bands of mineral alignment, which would appear as magnetic 'stripes' in the oceanic crust.

In 1963, Vine, along with his PhD adviser, Drummond Matthews, published a paper entitled 'Magnetic anomalies over oceanic ridges', which showed exactly that. The paper at last confirmed the reality of sea floor spreading.[11]

Resistance to continental drift was now futile. Professors whose entire careers had been based on old 'fixist' ideas had to finally

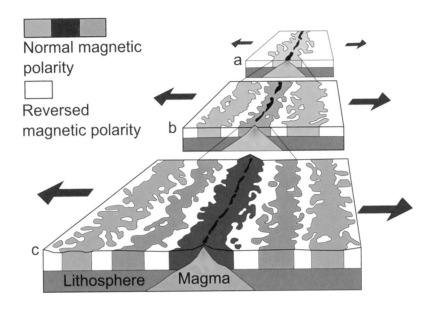

Normal magnetic polarity

Reversed magnetic polarity

a

b

c

Lithosphere    Magma

*Sea floor spreading reflected in 'stripes' of alternating polarity.*

abandon ship. For a younger generation of geologists, sea floor spreading would form the basis of their science from hereon.

The concept of plate tectonics was finally synthesized by the Canadian geophysicist John Tuzo Wilson, initially a 'fixist' but who had been swayed by the rapidly mounting evidence in favour of drift. Wilson came up with two groundbreaking concepts that would become central to plate tectonics. Firstly, island chains like the Hawaiian Islands, he proposed, were formed when moving oceanic crust passed over a 'hot spot' in the Earth's mantle. The volcanic activity in the Earth's crust periodically spewed lava onto the sea floor. Island chains were created as the sea floor moved away from the hot spot and new lava bubbled up in a fresh spot. Wilson's second major discovery was that of transform faults, which allow plates to move around each other without either being destroyed.

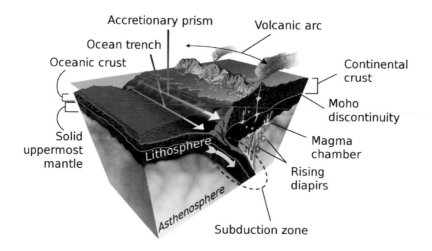

Accretionary prism
Volcanic arc
Ocean trench
Oceanic crust
Continental crust
Moho discontinuity
Solid uppermost mantle
Lithosphere
Magma chamber
Rising diapirs
Asthenosphere
Subduction zone

*A subduction zone at the edge of a continent.*

Wilson's discoveries unified the theory. With their publication, a world riven by plate tectonics was born, and the rewriting of Earth's story began in earnest.

# A RHYTHMIC MERGING OF THE CONTINENTS

In the beginning, around 4.5 billion years ago, there was fire, and the planet Earth formed from the accretion of material spewed out by a massive supernova, an exploding star.

Dust and gas congregated on the gathering weight of the fiery orb, then kilometre-sized (0.6 miles) hunks of debris. Finally huge, 100 kilometre-plus (62 miles) shards of rock were pulled into the inferno.

Churning in a vast ball of radioactive magma, the elements began to separate out, with heavier elements like iron and magnesium sinking deep into the melt, and lighter ones like silicon and oxygen floating to the surface.

The Earth slowly formed into layers — a solid iron core, a liquid outer core, a semi-solid mantle and, when the planet had finally cooled enough for rock to solidify, a hard crust which formed a complete 'lid' on the planet. This state persisted for hundreds of millions of years until ruptures formed in the lid and the Earth's crust began to fracture into tectonic plates. Plate tectonics could now begin.[12]

The crust sits on top of the slippery asthenosphere — a deep layer of hot rock which, although not molten, 'flows' in geological time — in the same way that old window-glass panes, which gradually thicken at the base over hundreds of years, can be said to flow. The semi-solid nature of the asthenosphere layer is what allows the more rigid plates to move across the top of it.

At mid-ocean ridges, oceanic crust is formed from liquid magma, spilling out of the Earth's interior and moving slowly away from the ridge. As it solidifies, it incorporates water into its molecular structure, carrying it with it on its long journey. As this waterlogged oceanic crust moves further away from the spreading ridges, it cools and sinks.

When it strikes the side of a continent, up to 180 million years later, it is subsumed, sliding down under the continent and deep into the mantle. Here, its rocks are converted by heat and pressure to new minerals. The descending crust takes all its water down with it too. It is estimated there is five times as much water bound up in oceanic crust than exists in the world's oceans.

As the crust is pulled down, this water triggers explosive melting in the mantle. Magma rushes to the surface to erupt as volcanoes. These volcanoes form a line along the front of the subducting plate — such 'subduction zone' volcanoes are South America's Andes, the volcanoes of North America's Pacific northwest, and the islands of Japan.

The subducting slab falls deep into the mantle, accumulating at the boundary of the mantle in a 'slab graveyard' before being chemically recycled into the Earth's mantle.

At the surface, extrusions of mainly silica-rich lava solidify to form continental crust, which is thicker but less dense than the surrounding oceanic crust. These continents 'float' on the underlying viscous mantle in the same way icebergs float on the sea. Their granitic roots extend deep into the mantle.

Plate tectonics constantly shift the continents around, splitting them apart and then bringing them back together. When continents clash, subduction is impossible and mountain ranges are instead thrown up. Periodically, all the continents on Earth are pushed together to form a supercontinent.

The movements of the continents are not random. They operate according to an internal rhythm set by plate tectonics. When a supercontinent forms, it thermally insulates the mantle beneath it, causing increasing temperatures under its mass. Eventually, this rise in temperature triggers a mantle superplume — a massive upwelling in the mantle, up to 3000 kilometres (1860 miles) tall, which billows up towards the crust at a geologically rapid pace. The plume strikes the base of the supercontinent, bubbling it upwards and outwards, fracturing it. Pieces of continent on either side of this fracture zone then begin to slide away on either side of it. Other forces also wrench these continental pieces — sideways forces at subduction zones and mid-ocean ridges that tear continents apart like a zipper.

After tens of millions of years of spreading apart, the oldest oceanic crust cools, becoming denser as it does. At this point, it begins to sink back into the mantle, reining in the outward movement. A new subduction zone and a new plate boundary are formed.

The switch is then flipped: the newly formed subduction zones begin hauling the continents back to where they came. Once again the continents are drawn back into their locked embrace. This is what's known as the supercontinent cycle, a repeated opening and closing of oceans, a rhythmic merging of the continents, like geological breathing.

Over enormous time scales — hundreds of millions of years — the supercontinents come and go: Ur, Kenorland, Columbia, Rodinia, Pangaea, Gondwana.[13]

This pulsing of the continents has profound effects on Earth systems. When the continents are heaped together, mountains and highlands are pushed up at the convergent plate boundaries. Having all this continental crust exposed to the elements at high altitude increases weathering, which through chemical reactions draws down carbon dioxide. Because carbon dioxide is a greenhouse gas that insulates the Earth's atmosphere, this drawdown cools the planet.

If the supercontinent stretches across the polar latitudes, ice caps form in the cooling climate. These have high albedo, which means they reflect more of the sun's energy back into space, contributing to further cooling. In such a configuration, the world slips into an 'icehouse state'.

When the continents begin to split apart, less weathering and erosion on land leads to a buildup of carbon dioxide — the Earth switches into 'greenhouse mode' and becomes warmer. The splitting apart of continents also fragments species, leading to increased speciation. New animals and plants arise rapidly and biodiversity skyrockets. In this way, the pulsing rhythms of the continents are intricately linked to the evolution of life.[14]

Plants, too, are shaped by these supercontinent cycles. But plants have not been passive agents in this planetary saga. Perhaps more so than any other kingdom of life, they have also shaped the Earth's climate and history, terraforming the planet for their own benefit.

# THE HOLY FIRE OF LIFE ON EARTH

Around 3.5 billion years ago, a spark was lit that would change the world. In chemical terms it looks like this: $6CO_2 + 6H_2O$ + Sunlight → $C_6H_{12}O_6 + 6O_2$

This equation — carbon dioxide plus water plus sunlight, giving rise to carbohydrates, with oxygen as a by-product — is the holy fire of life on Earth. It is the reaction that underlies the ability of every living organism to exist in this world, as almost all of us, from the tiniest bacteria to the biggest whale, rely upon it. We call it photosynthesis.

In simple terms, photosynthesis is the ability of plants and other organisms to manufacture their food from the sun. Cyanobacteria, microscopic creatures living in those ancient oceans, had the ability to photosynthesize from at least 3 billion years ago.[15]

The sun's potent energy, which until now was lost to space, was now captured by photosynthesizing organisms, which funnelled it into Earth's geochemical systems. As they did, they began changing the composition of the planet's atmosphere, stripping it of carbon dioxide. Photosynthesis splits the oxygen molecule into its constituent parts — hydrogen and water — with the oxygen molecule (the 'O' in $H_2O$) being liberated as a by-product.

Oxygen now flooded the oceans and atmosphere, profoundly altering life on Earth. This flood of oxygen provided the platform upon which complex life was able to evolve. Around 2 billion years ago, the first eukaryotic, or complex, cells appeared. In a twist that even the most imaginative science fiction writer might balk at, some of these early eukaryotic cells absorbed cyanobacteria, making them part of themselves. In return for room and board inside the cell, these organelles, called chloroplasts, went to work for the cells, taking on the role of photosynthesis for their mutual benefit.

The land was still a deadly place for life, in part due to the dangerous ultraviolet radiation that blasted its surface on a daily basis. As more oxygen was produced, however, oxygen molecules combined to form ozone in the upper levels of the atmosphere. This ozone layer provided a protective shield that blocked out the worst of the sun's deadly ultraviolet rays.[16]

And so photosynthesis laid the groundwork for the movement of life onto land. Gondwana was set for invasion, but getting there would be no easy feat. For organisms that had evolved in a watery world, the colonization of land was akin to venturing onto an alien planet.

Puʻu ʻŌʻō volcano in the Hawaiian Islands erupts lava into the Pacific Ocean.

# 470 MILLION YEARS AGO

Along the watery fringes of Gondwana, mats of green plant life only a few centimetres high — plants that look like lichens or mosses — arise from green algae that have lived in the warm seas around Gondwana for millions of years.

Like the algae they evolved from, these organisms reproduce using spores, female and male cells that meet in water to create the next generation. Because of this, these early land plants are tied to water.

These plants are not alone here on land. Fungi and bacteria have already inhabited Gondwana for millions of years and now they form an alliance with these plants. Fungi attach themselves to the plants' primitive root systems, helping them extract nitrogen and other elements from the brittle, impoverished ground in exchange for the sugars the plants create through photosynthesis.[17]

As soils become more developed, plants evolve deeper reaching, more complex roots, structures to bind them to Gondwana and draw out the scant nutrients in its rocky heart. Fungi and bacteria that live on these roots help them split the rocks apart, letting water in to erode them, releasing their life-giving minerals.

Competition for sunlight pushes plants to grow taller. They now need to transport water from their roots to their upper stems, and to transport the sugars produced by photosynthesis back down to the roots. To do so, they evolve vascular tissue — chains of open-ended cells called xylem and phloem, which act like veins and arteries in the plants' structure, carrying not blood but water and food.[18] They develop lignin in their stems, a polymer which gives them rigidity, allowing them to stand up straight on the surface of the Earth, reaching, as always, for the sun.[19] In places, plants now tower up to a metre (1 yard) above the ground.

Stomata, tiny holes on their primitive leaves and on the stems draw in carbon dioxide and expel oxygen. Vascular plants are living conduits between soil and sky. Gondwana breathes.

# 5

# WEEDS

For someone who writes about plants, my love of gardening has its limits. When I first bought my house in Port Chalmers its neat garden, set over three levels on the side of a hill overlooking Otago Harbour, was immaculately tended and manicured.

A succession of bulbs bloomed throughout the year, filling the garden with colour even in the winter months. Shrubs were neatly clipped and soil beds raked free of weeds. The lawn ended in well-maintained straight lines.

For the next few years (before my green-fingered wife arrived to set things straight), my busy life saw me apply a hands-off approach to gardening, and my backyard rapidly devolved into a tangled jungle that left me feeling woefully inadequate as I compared it to my neighbours' neatly tended backyards. Weeds charged up where cultivated blooms now struggled, racing for the light and throwing forth scrappy flowers that faded and withered in the sun before dropping off, leaving seed heads that dried and scattered their contents.

In corners that escaped my token efforts with the mower and the weed eater, dandelions, dock and wildflowers towered above a dense understorey of woolly grasses. Sticky sweet pea vines clambered up over everything until they formed a heavy mat that threatened to collapse in on itself. Voracious creepers wound their way through the branches and boughs of every tree and shrub they could get their tendrils into, sometimes strangling the life out of them as they filled their boughs with nefarious blooms. The prettiest perennials, so carefully favoured and nurtured by the previous owners, fell victim to tougher, wirier plants that had evolved to take advantage of their neighbours' weaknesses.

As I gazed out on the wreckage, I understood that, at its core, a garden is always imbued with the wildness from which it was cultivated. The jungle inhabits our ordered rows and neatly trained aspirations. It's written into the DNA of our ordered suburban lives.

Five hundred million years of hard scrabble for existence have scorched that fierce lust for life into every plant in our fields and gardens. Its expression is, to our eye, chaotic. It mocks our imagined sense of security and control, even though in the end, of course, we ourselves are really only a haircut away from that same primitive jungle.

# THROUGH A GLASS, DARKLY

The geological record is a vastly imperfect transcript of the story of life on Earth. It is fragmentary, distorted and broken. Huge parts of it are missing, and what remains requires highly experienced eyes to decipher it. Think of it as a book that has been shredded, burned and left out in the rain for a year or so. What scraps are left give us only a dim understanding of what has occurred — to borrow from the Bible, we see through a glass, darkly.

When plant material falls to the ground, most of it is consumed by fungi, bacteria and other detritivores. In certain situations, however, such as along the shores of shallow estuaries and in gently meandering river systems and lakes, some of this material may be preserved in layers of soft mud. Over time, leaves, stems and twigs get covered in silt and sand. As more sediment is laid down, these plant remains are compressed, undergoing physical and chemical changes until they are turned to stone. In places, their cellular structure is filled with minerals and 'permineralized', preserving the plant structure in fine detail. Captured in this way, plants can traverse geological time, emerging in our modern era as fossils.[1]

Around the world, there are a handful of localities where fossil plants are found in rich profusion and are preserved in exquisite detail — enough to reveal not just plant remains, but windows into ancient ecosystems. In these places, diverse assemblages of plants and the animals that lived among them are preserved together. These sites are known as lagerstätten, which in German means 'storage place'.

The names of the places they have been discovered have assumed legendary status in the world of paleobotany. Waterlogged environments tend to be best for preserving fossils, so many of the world's best-known plant fossil sites are in places where flooding has buried plant remains in mud or sand.

Occasionally, plants are also preserved in shallow marine environments — places where water pouring off the land deposits the material in sand or mud just offshore. One such site is in the Australian state of Victoria. Here, plants swept off the land in the late Silurian, 420 million years ago, were preserved in a shallow, muddy estuary that now outcrops as a crumbly shale on the outskirts of the town of Yea. Plant fossils were first discovered here in 1875 by William Baragwanath, but it wasn't until they were described by the

Baragwanathia *flora, Gondwana.*

famous Australian paleobotanist Isabel Cookson in 1936 that their true significance was recognized.[2]

The *Baragwanathia* flora, as it has become known, is among the oldest known collections of vascular plant fossils on the planet.[3] The rocks here record the earliest stages of plants' invasion of land, and may provide support to the theory that the first complex land plants evolved here, on Gondwana.[4]

The plants recorded at Yea are relatives of today's club mosses and extinct rhyniophytes — the very first lineages of early plants. *Baragwanathia longifolia*, the most common plant in this flora, is a thin, branching plant with tiny leaves held close to the stem in little fistfuls that cradle sporangia, the male reproductive parts of the plant.

The Yea site provides a compelling snapshot of the diminutive lives of plants in the Silurian and early Devonian, a time when plants were lowly margin dwellers restricted to the areas around waterways. But plants wouldn't remain as inoffensive bystanders to life on dry land. They were about to launch on a journey of immense proportions, one that would profoundly alter our world.

# 400 MILLION YEARS AGO

At last, Gondwana is coming into view. Descending from space onto a world incomprehensibly younger than the one we know, we see her for the first time — a vast bulk of land ballasting the globe on its southern side, encompassing a full eighth of the planet. From coast to coast at her widest point, she is 13,000 kilometres (8000 miles) across. To walk this distance would take the average person almost a year.

We descend upon Gondwana from the west, following the gleaming ribbon of what will one day be the Zambezi River far below, tracing in reverse its journey out of the Trans-Gondwana Mountains. Once, their peaks soared to the height of Mount Everest; now, after tens of millions of years of exposure to wind and rain, snow and ice, they have been ground down to relative stumps, crouched mountains that glower a dull red colour.[5]

Crossing this range we head south and east, flying over a vast, treeless land. The slopes of the hills and mountains are bare of plant life, brittle and dry. In between the ranges, big, braided rivers course out of the mountains. We follow one of these rivers as it leads us east, down and out of the central Gondwana highlands. Spreading away on both sides of the riverbed is a low carpet of green vegetation that troubles the feet of the barren hills but goes no higher.

After some time, we emerge out of the foothills and onto a wide floodplain. Lakes and streams gild the land, catching the sun as we pass overhead. Low plant life crowds around these waterways, the soft green valley floors exuding a veil of mist into the morning air. Monolithic *Prototaxites* fungi, some up to 5 metres (16½ feet) tall, stand sentry like primitive giant cacti at regular intervals across the landscape.[6]

At last, we touch down on the continent's northern shore, at the coast of what will one day be Saudi Arabia. Here, tall cliffs plunge into the Rheic Ocean. Gondwana vomits sediment into the ocean, staining the seas a muddy brown for several kilometres offshore. Unseen across this ocean, scattered fragments of old Rodinina — Laurentia, Avalonia, Baltica — collide to form Laurussia, the Old Red Continent, which converges on Gondwana, on trajectory for impact. Its inevitable collision is still many millions of years away.[7]

This ocean heaves with life — the sea floor scuttling with ancient creatures like trilobites, the water column stacked with the first bony fish. Sharks prowl these waters, too, ancestors of predators that will haunt the planet's seas in the time of humans 400 million years hence.

Walking the coastline, we traverse an inlet, a shallow, warm estuary that cuts into the side of the continent. Wading its fringes, the warm, salty water sloshing against our ankles, we emerge onto dry land again and find ourselves pushing through knee-high vegetation of small, stringy plants, their tough stems covered with tiny, spiky leaves. They make a little forest at our feet.

And now, as we watch, 40 million years rush by in a few minutes. For tens of millions of years, land plants remain fairly unchanged. But now, in response to the rapid proliferation of photosynthesizing plant life, carbon dioxide levels in the atmosphere are falling. Whereas the first plants had access to as much $CO_2$ as they could have ever wanted, now it is in shorter supply, so plants must evolve new technologies to gather more of it. Simple, bifurcating stems mesh to form broad platforms — solar panels — to capture more of the sun's energy. These plants hold their newly evolved broad leaves to the life-giving light.[8]

There is an explosion of plant diversity to rival the Cambrian explosion which occurred among animals in the sea millions of years earlier.[9] The vascular systems of these new plants now produce secondary material: cells of xylem, which form wood, and of phloem, which form bark. Wood facilitates more effective water transport,[10] and allows plants to grow tall, outcompeting their neighbours in the race for the sun. The first trees appear and suddenly forests spring up along the edges of rivers and lakes.[11]

These forests are dominated by a spore-producing tree called *Archaeopteris*, which looks like a large conifer with fern-like fronds. The intertwined root systems of these early forests bind soils, allowing, for the first time, meandering rivers to weave across the landscape.[12]

These rivers connect coastal wetlands with the continental interior, facilitating the movement of organisms inland. Leaf litter nourishes the streams, rivers and lakes they grow alongside, providing nutrients and habitat for freshwater fish, who explode in diversity in number.[13]

Egg-laying amphibians have swarmed into rivers, lakes and swamps, living and breeding in the shelter of wet woodland groves.[14]

The trees' huge roots run deep into the Gondwana soil, deeper than any other plant has yet managed to mine, ferreting out water from the ground and unlocking nutrients from the rocks and soil. Rain made acidic by carbon dioxide attacks this freshly exposed rock. The runoff draws its weathered minerals down into the sludge of the Gondwanan rivers. All of this pours into the ocean, carrying with it a torrent of silicon, phosphate, calcium and nitrogen. These nutrients fertilize the oceans and stimulate massive algal blooms which consume oxygen from the water, leaving dead zones where no life can flourish.[15]

Across the planet, huge numbers of marine species go extinct. More than 50 per cent of all marine life disappears.[16] And so plants bring down the curtain on the Devonian period. They've successfully taken over the world. What is to follow, the Carboniferous, will be a golden age for their kind.

# AGE OF THE MEGA-INSECTS

Adolphe-Théodore Brongniart, the great botanist who formally described and named *Glossopteris*, had a grandson, Charles, who followed in his grandfather, and great-grandfather Alexandre's, footsteps, taking a deep interest in fossils and the natural world.

Charles' primary interest was insects, and he became a pioneer in the field of paleoentymology, the study of prehistoric insects. For sixteen years he applied himself to the study of fossil-rich coal measures in Commentry, in central France. It was from here, in 1885, that he described the biggest insect ever known to exist on the planet. It was a huge dragonfly he named *Meganeura*[17] (which means large-nerved; the insect was named for the prominent veins in its wings), an enormous predator with a wingspan of 70 centimetres (27½ inches).

It was later suggested that *Meganeura* was only able to grow so big because it existed in an atmosphere that was thick with oxygen. Because of the way they respire through their bodies, insects have a natural limit to how big they can grow, and it has been argued that this limit is determined by oxygen levels. If oxygen levels were higher today, we might have dragonflies the size of small dogs, just as existed in the Carboniferous period.[18]

What made this world so oxygen-rich was the incredible forests that inhabited it. In the Carboniferous, huge swathes of lycopod forest, dominated by bizarre, scaly, spore-producing trees called *Lepidodendron*, cloaked the tropical regions of the world. On the northern continent of Laurussia, which at this time was situated on the equator, they dwelt in a world that was low-lying and swampy. They also existed, to a lesser extent, on southerly Gondwana.[19]

The biggest of these *Lepidodendron* trees towered 45 metres (148 feet) above the ground. They grew rapidly, reproducing just once before they died. In these forests, *Lepidodendron* dominated a rich understory of horsetails and ferns. In these whispering groves scuttled millipedes, springtails and mites, large amphibians and early reptiles.[20] To us, a Carboniferous forest would have seemed a very alien world.

*Trees of the Carboniferous Period.*

As a result of the massive amount of photosynthesis now occurring at the planet's surface, these forests were drowning in oxygen. At times, oxygen levels in the atmosphere were so high that trees were at high risk of wildfire.[21] Perhaps to combat this, the *Lepidodendron* trees drew up enormous amounts of water, enough to retard the flames of the conflagrations that would otherwise threaten them. Transpiring tons of water, they became like fire hoses pointed at the sky. The moisture they pumped into the atmosphere created the warm, wet microclimate in which they thrived.

With their spiky leaves, these trees strip-mined the sky for carbon dioxide, drawing huge amounts of it down into their twigs, branches, trunks and roots, sequestering it in the framework of their wood and carrying it into the ground when they died. An enormous amount of forest material, along with the carbon they contained, was swallowed, un-decomposed by this hot, swampy world.[22] In time it would form most of the Northern Hemisphere's major coal deposits.

It was the remains of these *Lepidodendron* forests that William Smith would go in search of as he began to unravel Britain's geology. As a result of this massive burial of organic matter, carbon dioxide levels plummeted, causing planetary cooling to go into overdrive.[23]

The glorious Carboniferous, a heyday for plants, was coming to an end. Gondwana was now firmly positioned over the South Pole, and the ice was creeping northward.

The Appalachians are the long belt of low mountains that runs for 2400 kilometres (1490 miles) down the eastern edge of the North America, from Canada to the state of Georgia. These forested hills,

*The Appalachian mountains, USA.*

strung with trails and streams, draped in fable and folklore, mark the scene of a supercontinent-sized wreck. Three hundred million years ago, during the Carboniferous, the Rheic Ocean that separated Laurussia and Gondwana finally closed up, drawing these two giant chunks of continent together like the jaws of a clamp.[24] This monstrous collision was the orogeny, or mountain-building event, that formed the Appalachians and completed the latest of the world's great supercontinents — Wegener's Pangaea.

Today, they might be pretty low as far as mountain ranges go, but in their infancy the Appalachians might have rivalled the Himalayas for size.

The evidence for this ancient collision is seen across the Northern Hemisphere in a swathe of mountain remnants that stretches across North America, western Africa and into the British Isles.[25] Over the hundreds of millions of years since this impact, the Appalachians, along with their European and African counterparts, have been ground down like old teeth.

The formation of Pangaea profoundly changed Earth's climate systems. The sheer vastness of the supercontinent created a 'mega-monsoon' climate, with violent swings between wet and dry seasons.[26] The interior of Pangaea became arid and dry, but also prone to monstrous storms that would periodically scour out its hills and valleys.

As the Earth cooled, ice caps sprawled across the southern polar regions. The cooling, drying climate was devastating for the lycopod forests that had evolved to thrive in warm, wet conditions. Across Pangaea, they were pushed back into smaller and smaller refuges.

Locked into their spore-producing ways, the great *Lepidodendron* trees that had dominated this chapter in Earth's history were not able to tolerate the aridity and cold, and the spectacular forests they had built over 50 million years of the Carboniferous period withered and shrank.

The Great Carboniferous Rainforest collapse was the most devastating plant extinction that has ever occurred on this planet. Along with the demise of the great rainforests, the fauna of huge insects that thrived in these forests also vanished, and many animals also went extinct.[27]

The passing of the warm, wet Carboniferous would usher in the cold, dry world of the Early Permian. For plants to survive this new age, they would need to be resilient. They would need to be able to handle long periods of desiccation, and they would need to be able to travel.

Fortunately, evolution had provided just the ticket.

# A LIFE CAPSULE

The seed is an incredible thing. In 2005, Russian researchers working in Siberia discovered the seeds of a small flowering plant called *Silene stenophylla* in a burrow excavated by an ice age squirrel 32,000 years earlier. Encased in permafrost, the seeds had been kept in a high state of preservation, so researchers decided to try the unthinkable.

They extracted embryos from the seeds, transferred them to vials and tried to germinate them. To their astonishment, it worked. The seeds sprouted plants that awoke to a laboratory spring in another age.[28] In human terms this was the equivalent of resurrecting a cave-dwelling Stone Age hunter and sitting him down for coffee in a modern city café.

That a seed should be able to do what it had evolved to do, 32,000 years after it was produced, is testament to the incredible power contained in one of nature's most effective contrivances.

Seed plants evolved from spore-producing plants. Instead of just firing off both male and female gametes into the water to do their own thing, these plants retained their female spores in an internal structure, where they were fertilized internally. The developing embryo grew in a protective covering — a seed coat.[29]

A seed is a life capsule for a plant embryo — a womb adrift from its parent body. The capsule is made of a hard, densely layered outer coat that protects the embryo from drying out and from the damaging effects of solar radiation. Once it finds itself in a suitable substrate and with enough moisture around, some of that moisture finds its way onto the seed, kickstarting the embryo into life. The seed coat splits open and, like a space traveller taking their first tentative steps on a new planet, the plant embryo puts out a tiny feeler root and reaches out for the Earth.

The feeler root is sensitive to gravity so is able to find its way down into the soil, where it immediately gets to work hunting for water. In trees, this feeler becomes a taproot that burrows deep into the Earth.[30]

Just as the development of the amniotic egg allowed animals to conquer the interior of the supercontinent, the evolution of the seed would likewise allow plants to survive and thrive in the rigours of the Permian.

With the evolution of seeds, plants were no longer tied to waterways. Now they could store their embryos for months and even years. Seed reproduction would allow plants to colonize the dry interior of Gondwana. Now, virtually nowhere was out of reach of the advancing plants. In place of the lycopod forests of the late Carboniferous, new forests would arise composed of trees and plants that had evolved to take advantage of this new technology.

The first true seed plants were plants now referred to by paleobotanists as 'seed ferns', a term applied to a wide group of plants that superficially resemble ferns but, unlike ferns, produced seed. Seed ferns proliferated across Gondwana from the late Carboniferous and into the Permian which followed it.

Chief among them was the tour guide of our tangled story — *Glossopteris*. It's time to go and meet her at last.

# 280 MILLION YEARS AGO

A *Glossopteris* seed germinates in the spongy soil of a Permian rainforest. Its coat cracks open, allowing the root to emerge, seeking, amid the loamy richness, water and nutrients to power the transformation written into its genetic material.

The fist of a young shoot breaks the surface of the ground, emerging into the shade of the forest. It bursts into a world of dripping profusion, a chattering whisper-world, cool and buggy. Dropping its protective embryonic leaf, it spirals towards the light.

Photoreceptors in its cellular structure are electrified by the brush of the sun. They trigger the release of a hormone called auxin, which causes the cells on the shady side of the plant to be longer than those on the sunny side. As a result, the young plant bends towards the light.[31]

Underground, roots squirm through the soil, branching out in search of water and minerals. Here, they meet fungi which also live in the soil, seeking union. The fungi send their hyphae deep into the cells of the roots, entangling themselves there. From now until the day the tree dies, the tree and the fungi will operate as one, the tree providing the fungi with the sugars and lipids it needs, the fungi providing the plant with water, and phosphorus and other nutrients from the soil.[32]

The *Glossopteris* tree grows towards a narrow gap in the forest canopy, stretching upwards and away from the ground, its vascular system humming with movement, water and minerals coursing around its body as it grows.

As the seasons come and go, the tree expands in rhythmic pulses, each season's growth widening the trunk, forming a new ring of xylem to add to its core.

At the canopy, the *Glossopteris* leaves lie oblique to the sun, their chloroplasts hungry for photons. Stomata open and close in pulse with the waxing and waning sun, allowing water to transpire and fill the forest air.[33] Later it will fall as rain, to run along streams and rivers that caress the tangle of *Glossopteris* roots. In the machinery of the leaves, the chloroplasts, little light-processing factories, are buzzing and humming, turning light into sugars that fuel the tree's growth.

Deep within the Permian soil, fungal networks expand, each fungal mycelium plugging into the root systems of other plants and trees. In this underground realm, a complex market exchange system thrives, trees providing their neighbours with what the other needs, each according to their needs, proteins and sugars passed from tree to tree along fungal networks.[34]

The forest breathes, and eats, as one organism.

# 6

# SURVIVAL COUNTRY

Argentina, 2018. Buenos Aires in the late spring, the heat and wind of the South American continent lambasting the city, ratcheting up the tension, throaty cars heaving like seals. I suck in the city till my lungs croak and wish for the cool of evening.

When it comes it brings rain, a drenching absolution that washes the city clean. In its cool aftermath I walk the streets in search of music, and eventually find it in a café where a quartet weaves jazz with tango — dark, diminished piano chords melding with the aching, creaking violin of the music this city gave birth to. I've got 2500 kilometres (1500 miles) ahead of me. Somewhere out there in this huge continent, Gondwana awaits.

In the morning, I rise early to meet Bari Cariglino, a paleobotanist with Bernardino Rivadavia Museum of Natural Science, and we pile into her Isuzu Bighorn. By daybreak we're on our way, riding the motorway south out of the city. Before long we're in a green, sweaty countryside where flocks of flamingos crowd lakes beside fields of soybeans and road signs warn of crossing capybara. The gateways to the estancias, or ranches, are ostentatious, each bearing the name of the spread in grandiose letters over the entrance. The homesteads are usually hidden out of sight, but occasionally I catch a glimpse of one, a grand mansion guarded by dense trees, homes made fat by the profits of wool and cropping in this rich farmland.

The miles disappear beneath our chassis as the country begins to dry out, becoming more sparse. Civilization is diluted by the spreading land until we find ourselves, startlingly, on a highway with no one else on it and no houses in sight. The oppression of the city begins to lift as the highway takes hold. At some point we cross a river where a sign advises we're now in Patagonia. Here, the towns begin to take on a sort of desolate bleakness that matches the dryness of the land, as if the country has sucked the juice out of them. Plastic bags dance on the wind and catch on cattle fences. All along the highway, roadside shrines spring up, little temples dedicated to a highway saint called Gauchito Gil.

Over the next two days, we make our way deeper into Argentina's south. Beyond Puerto Madryn, the estancia gateways have lost all their bravado — no grand entrances and tree-guarded mansions here. The wind batters tired, fading tin signs hung in nondescript gateways that punctuate endless fences. This is survival country, vast and deadly.

In the early 1800s, Welsh settlers, who had been sold a dream of a new life in a green and fertile agrarian paradise, arrived here and were shocked at what confronted them.[1]

Eventually, through sheer tenacity and with help from the indigenous Tehuelche people, the Welsh settlers managed to make a home for themselves out here on the steppe. To this day, there are towns in Patagonia where the Welsh language is still widely spoken.

In the early evening of our third day on the road, we arrive at a place called Fitz Roy, a sparse truck stop town named for the captain of the *Beagle*, the ship that famously carried Darwin around the world between 1831 and 1836. In 1834, Fitzroy and Darwin travelled up the Río Santa Cruz, which flows to the south of here, headed deep into the heart of Patagonia, to the edge of the Andes. Fitz Roy has little to it but a restaurant, a petrol station, a small grocery shop and a guesthouse. An Argentinian flag snaps in the breeze. Chickens scratch at a backyard in the shade of a broken wall where a mismatched scrag of kids hoof a soccer ball around a stony pitch. The town slumbers in the heat as the sparse highway traffic sweeps past.

The next morning we head off, farewelling the main highway to follow a lonely road into the steppe. A wind farm looms in the distance; 30 or 40 huge turbines seem to take an age to arrive across the flatness. We turn off at the gate to an estancia and trundle down a gravel track. Now we are bouncing across farm tracks, opening and closing wire gates, slipping along and between the folds of the landscape.

Guanaco, a relative of the alpaca, outnumber sheep out here, running from our approach in lanky herds. Rhea, the South American version of the emu or ostrich, also flee our noisy passage. Occasionally a mara, a bizarre Patagonian animal that is somewhere between a deer, a rabbit and a rodent, explodes out of the ground and hops away, flashing its white behind.

We've taken a wrong turn and rough tracks lead us deeper into the estancia. We are looking for a track to the river, but the road keeps sending us in the wrong direction. I get out to open gate after gate, wrestling with the homemade wire contraptions.

After a couple of hours we find ourselves at a dead end. Out in the middle of the estancia, a vast sky above and shapeless land everywhere. Bari decides it's best to turn back so we start trundling our way back along the same tracks. There is a sense of despondency in the vehicle. It's starting to look very much like the day will be a write-off. These roads seem a lot longer on the way back.

The land now takes on a more dangerous aspect to my eyes. I start wondering how much gas we have. We are well out of phone range. Finally Bari sees the track we should have taken first time through. She keeps driving until it disappears, fading into the landscape and then we're bouncing across a wild, trackless country.

Another gate. We drop into a deep depression in the land that suggests a river. But all of a sudden we're out of room to move — we halt at an incline too steep to negotiate. Bari stops the vehicle and we pile out. It's a relief to be out of the truck, to have the engine switched off and to at last feel the big silence of the place. Bari walks over to the edge of the incline to look for a track. I follow her and to my surprise begin to see what look like fragments of gigantic oyster shells scattered on the dry, dusty ground. I point them out to Bari and together we start fossicking around.

The outcrop we're standing on, it turns out, is chock-full of brachiopods, bivalves, corals, and bryozoans. Hundreds of kilometres from the sea, and in one of the driest places I've been, we're standing on the remains of a coral reef.

Bari bends down to pick something up. 'This,' she says, handing me a small fragment of something white and hard, 'is bone.'

We start looking around and quickly find more. Bari finds a huge vertebra. For the next two hours we scour the ground until we have collected a respectable pile of pieces. It's the remains of an ancient whale, perhaps 30 million years old. We also find shark's teeth,

*Paleobotanist Bari Cariglino at work in Patagonia.*

bright and sharp as if they fell out of the ancient predator's mouth only yesterday. Bari bags everything up and loads it all in the truck.

I'm finding it hard to believe we've stopped at this seemingly arbitrary spot and made such a momentous find. 'Welcome to Patagonia,' Bari tells me. 'Stuff like this happens all the time.'

It's an amazing discovery, but this place — a Miocene reef — is not the Gondwana I'm looking for. That place — the Gondwana of the Permian — is much older and is still out here to be found.

The following day, we set out again. Bari made some enquiries the previous night and is certain of the way this time, and after a couple of hours of trundling on meagre tracks we drop over an embankment and find ourselves in the valley created by the Río Deseado. We ford the river and after another half hour or so we arrive at the site, park the truck and climb up the side of the valley to the outcrop.

It's a nondescript chunk of rock protruding like the stump of a broken tooth out of the grassland. From up here, we can see out across the whole valley. Working to beat the heat of the day, we get to work with geological hammers on the rock. It breaks off in awkward chunks that must be split and split again to find the fossils. After several attempts, a blow from a hammer splits the rock at the point where the fossil lies, exposing the impression to air and light after 270 million years.

At first I'm only finding ferns, fossilized in a grey-green layer of muddy stone. Then, in the rock layer just a few centimetres above, I crack open my first *Glossopteris* fossil. Its unmistakable impression is delicately and finely preserved, and in the creases of the newly shattered rock I can see the tangled intricacies of its venations. The leaves lie stacked upon one another in messy nests, embedded in a fine-grained, greenish-grey mudstone.

I scrabble around in the rock, pulling apart its layers, exposing more and more of the delicate leaves. My hands break apart chunks of Gondwana. When these leaves detached from their parent tree and tumbled gently down into the waters of the swamp below, South America was still the western hip of the great supercontinent.

We could have driven a few hours east of here, to where the Río Deseado now spills into the Atlantic Ocean, and just kept going, driving across what is now Namibia on the west coast of Africa, and from there, many thousands of kilometres, into Antarctica and beyond. I could even have driven home to New Zealand (although I would have found it underwater).

I break open more fossils, finding *Glossopteris* leaf impressions everywhere now. I sniff the rock, fancying I can almost smell the mouldering vegetation in the water. This, then, is the Gondwana I sought — the Gondwana of the Permian. This narrow wedge of rock provides a tiny window through which I can glimpse and, if I close my eyes and imagine, smell, taste and hear it.

# 266 MILLION YEARS AGO

It's cold and wet. We push through a Mid Permian forest, brushing aside towering horsetails that flick us with their bottlebrush whorls of leaves, ducking beneath tree ferns that shower us with water from hanging fronds.

Alongside runs a shallow, swampy river, its waters stained brown by decaying vegetation. Weird amphibians scuttle away at our approach, diving into the water and snaking off through foamy rafts of tea-coloured scum. Large dragonflies brush the limpid surface before taking rest on the fronds of waterside vegetation.

There is something instantly strange to us about this forest. Suddenly it dawns: there are no birds here. The foliage rustles and shifts above us in an unnervingly empty way. But listen carefully and there are sounds amid the leaves — a low hum of insect life. If you look closely you can see them. The foliage is alive with mites and beetles, aphids, springtails, and cockroaches.[2]

Forming the forest canopy high above us are towering *Glossopteris* trees, their long, tongue-shaped leaves instantly recognizable. Scattered among these are smaller conifers called *Cordaites*, spindly trees with long, flat butter-knife leaves.[3]

The understory is of tree ferns and herbaceous species like club mosses and sphenophytes. The oldest of the *Glossopteris* trees are hung with epiphytes — perching ferns and other small plants that construct an aerial city high above the Earth, arboreal condominiums populated by fungi, insects and smaller plants.

In places, the big trees have tumbled and their long trunks slowly decay into the muck, feasted on by fungi and armies of rot-loving invertebrates. Our feet slough through mats of fallen leaves; it is early autumn and the deciduous *Glossopteris* trees are shedding their leaves in preparation for a long winter ahead. Already the forest canopy has taken on a rusty yellow-red hue. The leaves pile up in great drifts along the riverbanks, clogging up the muddy pools of still water where reptiles lurk and freshwater fish lay their eggs among the piles of decaying matter. The forest sighs with the wind.

But now, a far heavier sound shoulders its way through the boundaries of our perception. Crouching in the undergrowth so as not to be visible, we see the creature that makes its way towards us, sloshing through the shallow waters at the swamp edge. It's a lumbering animal that looks something like a Komodo dragon crossed with a hyena, its back angled downwards, its legs short, its body thick and bulky. At first glance it looks like some kind of reptile, but upon closer inspection, there is something more familiar, more mammalian about it. Its toothy head swings side to side as it emerges through the ferns that line the swamp's edge and lumbers onto dry land. It is a *Moschops*, one of the most common herbivores in this Mid Permian version of Earth.

*Moschops*' jaws are evolved to munch through dense vegetation, including the leaves and twigs of *Glossopteris*. We watch as the animal begins to browse. Suddenly, there is a commotion in the forest and three more *Moschops* emerge from the trees to join this first one. They move slowly through the undergrowth, their jaws grinding away at the *Glossopteris* foliage, grunting with satisfaction and bickering among themselves. Occasionally one lifts its tail to defecate, dumping a huge pile of brown-green faeces on the forest floor.

Two of the animals develop an unresolvable disagreement and square off to fight. They paw the ground and charge, lowering their heads at the last minute to absorb the blow of impact with their heavy, flat heads.[4] The thuds of their collisions reverberate through the forest. The fight, though, is half-hearted, and eventually both animals give up their argument and return to their browsing.

Then, one of the *Moschops* strays from the group and moves towards the water's edge, where it reaches for a horsetail that hangs low overhead. Its tugs against the bushy fronds release a shower of droplets across the still, calm water.

Suddenly the animal stiffens. The herd senses the change and their browsing immediately ceases. They, too, freeze. A half-chewed branch tumbles to the ground with a leafy crash.

The jungle explodes, the understorey parting to reveal the predator it concealed. It's an *Anteosaurus*, the dominant hunter of this forest. Its low, muscular body is hard as a spring, its jaws lined with rows of glistening teeth. Green eyes burn in the side of its head. Its body and being are taut with the hunt, jaws parted in brutal expectation.

The *Anteosaurus* lunges for the lone *Moschops*, which explodes into life, but too late. The hunter's powerful jaws lock onto its exposed neck. There is a tearing snort, a frothy tangle of jaws and eyes. The *Moschops* staggers beneath the weight of the attack and is drawn to its knees, head bowed.

The other *Moschops* are gone; in unison they crash off through the trees, branches crunching and snapping before their panic. At the water's edge, the *Anteosaurus* holds its victim in a death grip. All it needs to do now is wait for the big herbivore to tire.

The stricken *Moschops* is done for. Sensing its flagging strength, the *Anteosaurus* drags its victim down until the *Moschops*' nostrils are submerged in the swampy mire. The beleaguered animal snorts a few times, the water churning in bubbles of bloody spray. Bit by bit, and gradually, it succumbs, drowning in the muck. After some time, the *Anteosaurus* hauls the *Moschops*' limp body out of the water and drags it into the forest to devour it.[5]

Silence returns again to the glade. The *Glossopteris* forest returns to its long dreaming.

For more than a century, scientists like Bari Cariglino have wandered the broken fragments of old Gondwana in search of fossilized remnants of these *Glossopteris* forests, looking for the secrets of life in the Permian they might reveal. Their work has taken them from the coal-mining heartland of India's northeast to the barely explored trans-Antarctic mountains, where remnants of *Glossopteris* tumble out of frozen rock into an ice-filled environment so unlike the one they grew in.

These scientists go to work on the velds of Africa, the steppe of South America, in the outback of Queensland and the jungles of Brazil. In looking for *Glossopteris* they, like me, are searching for remnants of that long-lost great southern land — Gondwana — seeking to understand it, to walk among its forests and sense its changing seasons. In doing so, they try to understand more about the world we live in and to better understand humanity. Because that ancient Gondwana forest is part of us: we are connected to it, through every cell in our body.

Through the great tree of life every bird, fish, mammal, reptile, plant, fungi and bacteria on the planet can trace its ancestry back to a common ancestor. This was the great insight glimpsed by Charles Darwin and Alfred Wallace, and the fundamental understanding that informs all of biology today. Every tree and shrub that grows in that *Glossopteris* forest, every reaching vine and clinging fern, and every animal that moves among it is, to some extent, our relative.

The job of sorting out where each species in the history of life sits in relation to one another falls to taxonomists. Looking for those common ancestors and tracing their lineages is an unceasing toil, and unravelling the family dynamics of a genus of plants that has been extinct for 250 million years is not easy.

*Glossopteris* is mostly found as fossils of detached leaves — vast swathes of them. Rarely are the leaves found attached to twigs or branches. As a result, no one really knows what a complete *Glossopteris* tree actually looked like. While it's fairly certain that most *Glossopteris* species were large trees, it can't be ruled out that some shrubs and smaller plants also existed.

After almost two centuries of study, there are now over 100 described species of *Glossopteris*, and many are controversial.[6] It is generally assumed that trunks found in association with layers of *Glossopteris* leaves are of *Glossopteris* trees. This wood is assigned to a 'form-genus', *Araucarioxylon*, being similar to the wood of modern *Aracauria* trees, such as monkey puzzles.[7] The trunks reached at least 80 centimetres (31½ inches) in diameter, indicating the trees may have grown to around 30 metres (98 feet) in height. The leaves were borne on branches in tight, dense bunches or on short shoots similar to ginkgos.

The unique root systems of *Glossopteris* plants also have their own scientific name: *Verterbraria*.[8] These root structures are easily identifiable by a jointed, chambered structure which is thought to have allowed the roots to aerate in a wet environment. The roots were shallow and woody. From examining these roots, scientists have concluded that *Glossopteris* was a swamp-loving 'mire' plant, best suited to low-lying moist environments in the cool, wet climate of the Permian temperate regions. The sedimentary rocks *Glossopteris* is found in are usually the remains of winding rivers and lakes, confirming that these trees thrived in wet conditions.[9]

It is likely many, but probably not all, *Glossopteris* species were deciduous, losing their leaves every autumn.[10] Their bark is thought to have been thick and stringy, much like a *Eucalyptus*.[11] Also like

modern-day *Eucalyptus* species, *Glossopteris* had epicormic shoots, buds which lie dormant under the tree's bark, only springing to life when light levels in the forest are suddenly increased, such as when a taller neighbouring tree falls or when the tree is exposed to sudden stress. These shoots, which on *Eucalyptus* trees throw out bunches of bright, fresh foliage after a forest fire, would have allowed *Glossopteris* to immediately rebound after wildfire swept through the forest.[12] Like the stringy bark, epicormic shoots are likely to have been an adaptation for a time of high fire risk. Charcoal deposits indicate wildfires were common in the Permian.[13]

The first paleobotanist to describe the reproductive parts, or fructifications, of *Glossopteris* was the South African paleobotanist Edna Plumstead. In a manner unlike any modern trees, some *Glossopteris* leaves had evolved to carry the fertile organs. The female ovules were held in large bunches on foliage leaves, while the smaller male structures hung from the leaves like tiny trinkets. They come in a bewildering array of shapes and sizes. On the basis of these fructifications, Plumstead concluded that *Glossopteris* belonged in the family of plants called gymnosperms, the group that includes all modern conifers. The *Glossopteris* fructifications came in many shapes and sizes — this is part of the reason we know there were so many different species of *Glossopteris*.

Paleobotanists are able to map out the *Glossopteris* family and trace its evolution through time by looking at these structures. Over time, the fructifications gradually became more closely fused to the leaves that held them, until the cover leaves that protected them curled around the fruit in a way that suggests an early form of flower. These alterations are, in part, what has led some researchers to consider that *Glossopteris* might be the ancestor of modern flowering plants.[14]

*Glossopteris* male sporangia produced pollen, probably in huge amounts. This pollen would have been distributed across the forest by wind, with just a miniscule fraction of it finding its way to a receptive female appendage. When the pollen reached a female fructification, it travelled down the pollen tube using a tiny, mobile sperm, in the manner of modern ginkgos and cycads.[15]

*Glossopteris* trees first appear in the fossil record in the Early Permian, a time when the Earth was beginning to warm up again after the brutal cold snap that ended the Carboniferous period. They came into being as those late Carboniferous ice sheets retreated, living alongside the retreating glaciers and following them south as they receded with the warming climate.

From its earliest times then, *Glossopteris* was capable of thriving in a cold, tough environment. Perhaps it was these early trials and tribulations that prepared it for its great success as the age proceeded.

As the Permian progressed, *Glossopteris* spread across temperate Gondwana, dominating the vegetation, especially in the low, wet areas of the supercontinent. Despite the harsh, dry conditions that dominated much of Gondwana, *Glossopteris* species were found across most of the continent, from its moist and temperate coast to its arid heart.[16] They formed the Southern Hemisphere's first great seed plant forests. They were one of the most successful plant groups ever to grow on the Earth.

And were we to venture back in time to the Mid Permian, there is one particular aspect of *Glossopteris* life that would utterly confound us, as we have nothing in our modern world to compare it to. In the Permian Gondwana, there were forests at the South Pole.[17]

# 260 MILLION YEARS AGO

We are southbound, drifting across the interior of southern Gondwana. Below us, a *Glossopteris* forest stretches as far as the eye can see. Here in the polar regions a night-free summer lasts for three months, before the sun disappears below the horizon for much of winter, plunging the land into long winter darkness.

It seems incredible that trees, which depend on the sun for their food, can survive under such conditions. And yet here in this Mid Permian Gondwana, they somehow do.

Being deciduous probably helps — the *Glossopteris* trees can shed their leaves and essentially shut down their life-support systems, effectively going into hibernation for the duration of the darkness. It helps, too, that this world is relatively warm — the pole-to-pole configuration of the land permits tropical and polar waters to mix, creating a much lessened temperature gradient between these two regions.

When the seasons change, many *Glossopteris* species rapidly shed their leaves, shutting down their biological systems to wait out the cold, dark months of winter. For three months the boreal *Glossopteris* forest stands in utter silence, towering trees standing sentinel over a planet on hiatus. This is a survival forest. The trees are thin and tall, with cone-shaped crowns evolved to best intercept the fragmentary light in these high latitudes. They crowd together, just metres apart, as if huddling together for warmth.

With the return of spring, the silent forest awakes. Insects are suddenly everywhere, rapaciously devouring *Glossopteris* leaves and laying eggs on their undersides. Fungi thrive in this moist environment and inhabit the very structure of the forest, burrowing their hyphae into the soil and rotten wood and establishing themselves on the trunks of trees and in the cradles formed by epiphytes.[18]

We pass over 75°S, the point where in 260 million years the *Terra Nova* expedition will encounter the edge of the Ross Ice Shelf. Here in the Permian, it is a low line of forested hills.

We pass the 77th line of latitude, where in 1911, Captain Scott will establish his base camp.

We keep going, travelling across the boreal expanse of what will one day be Antarctica. At last we touch down in a cool, secluded grove at the edge of a tiny lake, the low light of the late summer sun dappling the leafy ground beneath a towering *Glossopteris* tree. A boisterous breeze breaks out in the canopy high above, shaking the bunched foliage of the *Glossopteris* fronds. A single leaf detaches and flutters down through the forest and, as we watch, comes to rest on the surface of the lake. For a few minutes, it sits there. A small fish rises to the surface to investigate it but, finding nothing edible about it, retreats to the murk of the pond. The leaf is pushed at by the gentlest breath of air, and moves up against the bank. It becomes waterlogged and sinks a few centimetres until it comes to rest on the muddy bottom amid a tangle of fallen twigs. It does not move again.

At this exact spot, in 260 million years, Captain Robert Falcon Scott will break the fossilized impression of that leaf from the Beacon Sandstone, now exposed in a rocky outcrop amid snow and ice, load it into his canvas bag and lash it down to the team's dangerously heavy sled. He will carry it to his snowy grave.

Glossopteris *fossil, Falkland Islands.*

Clouds build up over the Paleo-Tethys Sea, the body of water that splits Pangaea in two, sucking in huge amounts of water and heat from the surface of the warm ocean. The stored energy of the weather cell builds and builds, ionizing the atmosphere and electrifying the sky.

An enormous low-pressure system over Gondwana is drawing air away from northern Pangaea, pulling the world's weather southward across the Tethys. The vast continental expanse to the south, heated by the remorseless sun, provides the engine for this Pangaean megamonsoon.[19]

Lightning rips the jet-black clouds above. The storm builds on its core of violence, drifting across the sea, drawing up more and more moisture and growing uncontrollably. It heads southwest with malevolent languidness.

In the dead of night, the storm makes landfall along the coast of northern Gondwana and unleashes its fury. The towering *Glossopteris* forest at the coast is shaken and torn, its canopy rippling and shuddering with the weight of the wind. There is a horrendous screaming of shredded vegetation.

Torrential rain pummels the forest and rivers swell to enormous torrents, escaping their banks and ripping through the bases of the tree trunks, tearing tons of vegetation from the Earth. Whole *Glossopteris* trees are torn from the swamps and cast to the ground, their root systems intact and exposed to the sky like giant claws, their connection to the land forever broken.

The storm tears inland, leaving in its wake a forest stripped bare and shattered. River basins are flooded from side to side.

Hundreds of kilometres from the coast, the land has largely absorbed the might of the weather system, and it has significantly weakened.

Finally, the storm blows itself out. With dawn, sunlight returns to the coast and a fierce sun burns unrelentingly across a vivid blue sky.

It will take months, even years, for the forest to recover, and yet recover it will. Along this seaboard swathe of Gondwana big storms regularly sweep in off the sea, dumping their consignment of rain along the bands of low hills that line the coast. As these storms continue inland, empty now of their rain, they become dry winds that suck moisture from the land. They desiccate the continent, creating a searing aridity in the interior. Far from the ocean, the whole interior of Gondwana would in fact be desert were it not for the sprawling *Glossopteris* forest. The forest absorbs rain in the coastal band, taking it in through its roots, using it in its photosynthesis and then transpiring it out through its stomata. This exhalation creates clouds above the forest, the trees providing a cool, moist habitat for themselves.

Roiling clouds of the moisture are carried further inland by the wind, where they once again deliver their load of water as rain to inland forests. In this way, the *Glossopteris* forests carry water from the coast into the heart of Gondwana, the trees holding the water aloft like an aerial river.[20] It is, in part, this system of constant hydration that allows life to flourish deep in the interior of the great continent. In the nurturing embrace of these forests, ecosystems evolve, every species part of the gathering complexity.

The animals of Gondwana — the synapsids and sauropsids — diversify, evolving over millions of years into a bewildering array of creatures, each of which finds their niche in this world. In the Mid Permian, the dinocephalians — a group of therapsids that includes *Moschops* and *Anteosaurus* — dominate Gondwana life. Then, at the end of the Mid Permian a changing climate, perhaps triggered by volcanic eruptions in the Northern Hemisphere, creates a devastating ecosystem turnover on Gondwana.[21] The dinocephalians disappear. Plants too, are hit hard by this event. *Glossopteris*, however, survives and continues to thrive in a world that is gradually getting warmer and drier.

Now, new lineages of therapsids, including cynodonts and gorgonopsids come to prominence, filling out many of the old niches the vanished dinocephalians have vacated. Herbivores grow big and powerful, and so the glossopterids develop defences against the herbivores' grazing — toxins in their cells and smaller, less palatable leaves.[22] Herbivores, in turn, develop more effective browsing jaws. They mow their way through enormous quantities of *Glossopteris* leaves, depositing the digested remains back in the soil in the form of dung that fertilizes the understorey, nourishing the forest. Their predators grow faster and more powerful in order to tackle them. Everything maintains an uneasy sense of balance — a cold war calm.

Beneath it all, Gondwana stirs ominously. The old agitations of its tectonic destiny don't ever sleep.

# 7
# THE OCCUPATION

An all-night bus trip through Patagonia, a long sleepless vigil staring out at the passing blackness of old Gondwana. Racing through silent, nameless midnight towns, looking out through the bus window at shuttered streets and sleeping houses, wondering about the lives that exist here that I will never know.

Now it's 5 a.m. and I'm stranded in the Río Gallegos bus station, awaiting my connection through to Punta Arenas. I watch the industrial outskirts of this grimy southern Patagonian city come to life through the windows of the bus station.

I try to make myself comfortable on the hard plastic seats of the bus station and watch the clock move. Anyone who says it is impossible to truly understand geological time should experience some of it in the Río Gallegos bus station.

The relief when my next bus finally arrives is enormous. The wild sense of movement after a long stasis, the spirit lifting as the grime of the town recedes and we move west towards the cool of the mountains.

There is more waiting in store, however. Crossing into Chile is not an easy process — the two countries are at a near-constant state of animosity, and getting through the border is literally a military operation.

From here, the road slides away more easily beneath our wheels. That night, I'm in Punta Arenas, my jumping-off point for my next Gondwana stop.

Punta Arenas sits on the Straits of Magellan, breathing the cool, foggy air that rolls in off the Southern Ocean. It is leafy and pleasant, a soothing contrast to the hot, dry, dusty steppe towns I have just emerged from. I have a day to fill here, so I head up to wander around the stand of *Nothofagus* beech forest that clothes the hills behind the town. This forest feels very familiar. *Nothofagus* species have a southern-wide distribution, being found in southern Australia, Papua New Guinea and here in South America. It is also the dominant forest type in my homeland, the South Island of New Zealand.

*Nothofagus* vaguely inhabits the same kind of austral latitudes *Glossopteris* once did, and in that sense is a modern analogue, although this is not really accurate, as where *Glossopteris* was a lover of wetlands, *Nothofagus* is a genus of the hill and mountain country. Still, walking in this Chilean forest is a reminder of just how closely the flora of our two countries are connected, despite the vast ocean that now separates us.

But Punta Arenas is just a jumping-off point for me. I'm headed for a slice of Gondwana that has had a very unusual journey indeed. By the middle of the next day I am pressed to the Perspex window of a Boeing 767, looking down at the barren, wave-hounded expanse of the Falkland Islands, looming up at me from amid the Atlantic Ocean.

# THE FALKLANDS

Early on the morning of 2 April 1982, Argentinian army forces came ashore at the Falkland Islands port of Stanley. There they encountered a fierce, albeit futile, resistance from a small force of Royal Marines that had been expecting them. Within fairly short order the Argentine forces were able to occupy the town.[1]

Over the coming days and weeks, Falkland Islanders huddled behind closed curtains in their homes, getting used to the idea that their quiet, sheep-dotted islands were now under occupation by a hostile military power.

The invasion sparked a fierce response from Margaret Thatcher's Conservative British government, which immediately dispatched a force to the South Atlantic to seize back control of its dependency. Britain was now, unofficially, at war with Argentina.

The Falklands conflict lasted for ten brutal weeks. It claimed the lives of 900 soldiers and sailors, mostly Argentinian recruits who were young, poorly trained and had no real idea why they were even there in the first place.

The reasons for the invasion remain muddy to this day. In part, it was probably an attempt by Argentina's military leaders to deflect attention from the barbaric decline of their regime. (Up to 30,000 Argentines, including many dissidents, students and intellectuals, had disappeared under the military junta's state-sponsored 'Dirty War' since 1976.) In order to stir up national fervour, the junta seized on a popular target. For generations Argentines had felt the Falkland Islands, or Las Malvinas as Argentines know them, belonged to them. Geographically, this would seem to make sense — after all, the Falklands lie just over 480 kilometres (300 miles) off the coast of Argentina. Intuitively, the two landmasses would seem to be connected. And yet, in the course of deep time, no such rules of obviousness apply.

It was Alexander du Toit who, in his 1937 book *Our Wandering Continents*, first made a remarkable observation about the Falkland Islands. Geologically, their rock strata have almost nothing in common with that of neighbouring Patagonia. Instead, they clearly and exactly match the rock strata of southern Africa — the Gondwanan Karoo sequences — 6500 kilometres (4000 miles) away.[2]

In 1952, the South African geologist Ray Adie took Du Toit's insight one step further. He realized that the geology of the Falkland Islands fits like a jigsaw puzzle not with western Africa, as a glance at a map of the world might suggest, but with eastern South Africa — the opposite side of the country from which the Falklands currently lie. Adie's conclusion, which has subsequently been reaffirmed by geologists, was that when Gondwana was assembled, and South America and Africa were joined as one, the Falkland Islands were wedged in between southern Africa and Madagascar. As the South American Plate swung away from the African Plate, it took the Falklands with it, rotating the islands 180 degrees until their north became their south.[3]

Armed with little more than a few hours of paleobotanical experience, I have come here to go looking for *Glossopteris* — to go looking for Gondwana. My flight lands at the Falklands' international airport, which is hosted by the huge military base on the island.

Our 767 taxis in among military troop carriers and fighter jets. Around 2000 British soldiers are stationed here. They use the Falklands as a training ground, but also to maintain a state of readiness in the highly unlikely case that Argentina should ever try to reinvade.

I make my way through the small customs desk and bundle my gear into a rusty Land Rover with a cracked windshield. Most of the trucks on this island, I soon learn, have cracked windshields — a result of the unpaved gravel roads spread with hard quartz pebbles. My reaction to the Falkland Islands is initially similar to that of Charles Darwin two centuries earlier: I find myself daunted by the naked hostility of the land, its confronting treelessness, its harsh, quartz-edged exterior.

In the town of Stanley, I stay in a bed and breakfast run by an elderly lady called Kay McCullum, who cooks me meals each night and regales me with stories of the town's history.

For several days, I explore the immediate area, climbing into the hills behind Stanley (a battlefield during the Falkland war, where I find the rusted remains of an Argentinian camp kitchen and watch a pair of falcons nesting on a craggy cliff). Wandering around Stanley, I eat egg rolls and drink coffee. The town has everything you might expect from a British seaside town: cosy pubs serving English stouts and ales; a night club where drag queens (yes, drag queens) mingle with heavily intoxicated soldiers from the military base, and a wharfside caravan serving fish and chips. Everywhere there are memories of the war, and lingering hostility towards Argentina still festers.

The big talking point on the island when I arrive is the proposed establishment of a new flight link between Brazil and the Falklands which, because it crosses Argentinian air space, is required by the Argentinian government to include a stop-off in the city of Córdoba. Many on the island are vehemently opposed to the new flight path

— they see it as a concession to a hostile country that still wants to own them.

My target here on the Falklands is the southern portion of East Falkland, an area known as Lafonia, after the Lafone brothers, traders who lived and worked here during the islands' mid-nineteenth century days as a remote cattle-ranching outpost.[4] Lafonia is one of the least-visited parts of the Falklands. Lacking the dramatic scenery of West Falkland and removed from most of the best wildlife-spotting sites, few tourists make the trek there. My desire to go there is therefore something of an oddity to Kay. She tells me that in her 70-odd years of living on the Falkland Islands, she has only made the 90-minute drive down to Lafonia once, and she didn't like it. It is, she tells me, boring. Kay is a woman of fine hospitality, many garden gnomes and very firm opinions.

Despite Kay's misgivings, I hire a four-wheel drive and head off on the road out of Stanley. After an hour on gravel roads, I reach the tiny settlement of Goose Green, a collection of red-roofed farm buildings sitting on the isthmus between the northern part of East Falkland and Lafonia. Goose Green was the site of an Argentinian air base and of one of the fiercest battles of the war, during which British amphibious troops landed and took control of the isthmus. Today it's a peaceful spot; there seems to be no one around when I arrive. A few sheep graze in a nearby paddock and a flock of hens pecks in the dust. One of the buildings has a little cafeteria in it, and here I pull in for a cup of tea and a sandwich before heading south.

In crossing the narrow isthmus of land upon which Goose Green sits, I enter a new geologic province, leaving behind the jagged quartzite hills of the northern part of the island and arriving into Permian Lafonia. Instantly, the landscape changes, reflecting this change in the underlying geology. Lafonia is the remains of the bed of a huge lake that sat in the heart of Gondwana during the Permian. It is this geology that makes it almost unnervingly flat. I find the flatness of the region — which Kay had found so boring — intriguing.

The land seems to vanish in all directions. Silence sits heavily on the sun-rusted grasslands. Vultures pivot on their wingtips in the air high above. Angling across the navel of the Falklands towards the coast, I pull over to breathe in the land, inhaling its dry grassiness, warming my skin against the expanse of the sky, absorbing the quiet. It feels good to be out here, far from the bustle of Stanley.

It was on Speedwell Island, just off the coast of Lafonia, that Johan Gunnar Andersson, of the Swedish Antarctic expedition of 1901–1903, discovered fossils of *Phyllotheca* and realized that the Falkland Islands had once maintained a connection with the Gondwanan continents.[5] In 1907, another Swedish botanist, Thore Gustaf Halle, travelled to the Falkland Islands and, inspired by Andersson's discovery, went in search of more convincing evidence. He found *Glossopteris* fossils in several locations on Lafonia.[6]

I camp in a farm settlement called North Arm, and in the morning wander across paddocks to the edge of the land, creeping along the edge of a narrow sea inlet with my geological hammer, stopping here and there to claw, jimmy and smash pieces of rock from the exposed outcrops the sea has made. Wherever I swing my hammer, I find *Glossopteris*, the unmistakable leaves encased in the dark mudstone that once formed the bed of that ancient lake.

In the afternoon, the wind riles in off the Atlantic, bringing with it a squadron of giant petrels, ocean wanderers. They climb the face of the wind and hang motionless above the land and the estuary, as if suspended in time.

For two days I explore Lafonia, stopping at known fossil sites, finding *Glossopteris* specimens at each one. In one place, I even find a chunk of an ancient tree trunk. I load them all into my truck and head back to Stanley. Since Halle, very few people have collected *Glossopteris* at the Falkland Islands, so I feel some sense of achievement when I bring my box of fossils into the local museum, label them up as best as I can and leave them there for safekeeping.

My time in the Falklands comes to end, and in the shimmering light of a wild and windy South Atlantic day I rise in a silver dart from the Falklands airport, on the first leg of my journey home to New Zealand. In the light of a landless dawn, I gaze down on the Atlantic's rippling expanse and marvel at the thought that when those plant fossils I found in Lafonia grew, this ocean didn't exist. In the time between them living as leaves on a tree and me digging them as fossils from the Falkland Islands rock, the Atlantic has ever so gradually opened up to form the vast gulf it does today. I can think of few better examples of the slow creep of geologic time than this: the existence of a great ocean where once there was once no water at all.

*Mount Harriet, Falkland Islands.*

# 254 MILLION YEARS AGO

And now we are floating high above the Gondwana of the Late Permian, breathing pristine air that has never tasted the soot of a human city or the stench of a jet plane's exhaust.

This version of the world is one massive continent and one ocean — Pangaea sprawling in a gigantic C-shape across the Earth, from pole to pole, incized at the middle by the Tethys Ocean, which cuts out the middle of Pangaea like a wedge taken from an orange. Gondwana ballasts the south, the bulk of her land below the 50th parallel. Surrounding all of this, the Panthalassic Ocean.[7]

These Permian seas are rich with what, to us, would appear an old, strange fauna. Giant swimming molluscs called ammonites are present, as are trilobites, armoured denizens of ancient seas. Rugose and tabulate corals, utterly unlike corals we are familiar with today, form the basis of the marine ecosystems along the continental shelves where warm, shallow seas lap the edge of the land. These seas heave with unusual fish.[8]

We approach Gondwana from the northwest, drifting over tropical forests dominated by ginkgos and conifers at the equator. As we move south, the forests dry out and the land becomes arid and dry. Crossing the part of the country that will one day be part of Niger, in Africa, we encounter a harsh, hot land periodically slammed by monsoon rains that fill the lakes and rivers that bejewel the country.

Sauropsids, the ancestors of reptiles, have evolved throughout the Permian into a wide array of species and in their many forms they now proliferate here. Among the most impressive of them are the pareiasaurs — the 'Blinkwater monsters' of Andrew Geddes Bain — heavy-footed tanks of animals that pug the Earth around the edge of these lakes, chomping through the tough woody plants with their industrial-strength jaws. Smaller amphibians and strange-looking fish lurk in the waterways. Aquatic lizard-like reptiles called captorhinids graze the edges of lakes and streams, feeding on the horsetails and other plants that live there.

Drifting south, we float above a vast, hot desert, stretching off in every direction as far as we can see. Temperatures here regularly soar above 50°C (122°F), making life almost unbearable. However, after a time, as we get further south, the desert begins to cool and thicken with plant life.

High above the country that will one day form Zambia, we cross woodland threaded by meandering rivers and gleaming with lakes. The valley floors along the edges of the waterways are crowded by a dense *Glossopteris* forest, the crowns of the biggest trees emerging from the canopy like broccoli heads. Horsetails, sphenophytes and ferns fill out the understorey. The hillsides above the valley floor are shaded by conifers and ginkgos.

We head south and west, heading towards the coast, flying high above the margin of Gondwana. Below us, the ocean thunders against the shores of a mountainous country punctuated by volcanoes that form a seemingly endless line right along the base of the supercontinent. These are the Gondwanides, a cordillera stretched across all the landmasses that will one day form Africa, South America, Antarctica and Australia. Along one part of the cordillera, in the region of western Gondwana that will one day be southern Africa, the mountains have risen to such great bulk that they have depressed the Earth's crust, leaving a wide, shallow basin into which big rivers drain.

In this harsh, arid land we encounter wandering herds of dicynodonts, tusked, warm-blooded animals that look like a cross between a tapir and a prairie dog — Bain's 'bidentals' in the flesh. They are browsers, moving among the shrubland and scraggly *Glossopteris* woodland, grinding away at the foliage with machine-like jaws. On flat lands they move in herds among pastures of jointed *Phyllotheca*.[9] In scattering ambush, they are run to ground and preyed upon by ferocious, sabre-toothed therapsids called gorgonopsids, the sleek killers of the Permian savannah.[10]

We cross the remains of a shallow inland sea, where fossils of the extinct Early Permian *Mesosaurus* erode out of limestone bluffs, and enter the part of Gondwana that will one day form Antarctica. Here, we turn north again, following an enormous river — perhaps the longest that has ever existed on Earth, pouring out of the Gondwanides and draining, as most water does here on Gondwana, northward.[11]

The river leads us again along the edge of a large inland range, through the arid highlands of Central Gondwana. There are no grasses in this Permian world, so the ground is instead dotted with small herbaceous plants — ferns, sphenophytes and tiny lycopods. Now, we near the coast of the Tethys Ocean, in the region of a future India. Here, moisture from the sea returns a lush verdancy to the temperate land and once again we find dense forests of the *Glossopteris* flora confounding the twisted waterways.

Far across the Tethys, the landmasses that form the Cimmerian terranes, parts of land that will one day form Turkey, Iran and the Malay Peninsula, are drawing away from Gondwana, hauled away by a slowly spreading sea floor that is also tugging at the edges of the supercontinent, thinning and stretching it. This stretching has created basins into which sediments from surrounding hills and mountains pool over millions of years. Meandering rivers wander at will across a wide flood plain, periodically burying the debris of the profuse *Glossopteris* forests that line their banks.[12]

Whole *Glossopteris* forests flourish, then fall and are buried. Massive lenses of organic matter accumulate in the Earth, slowly turning to peat, then coal, destined to be dug from the Earth in a distant time, a time very unlike this one.

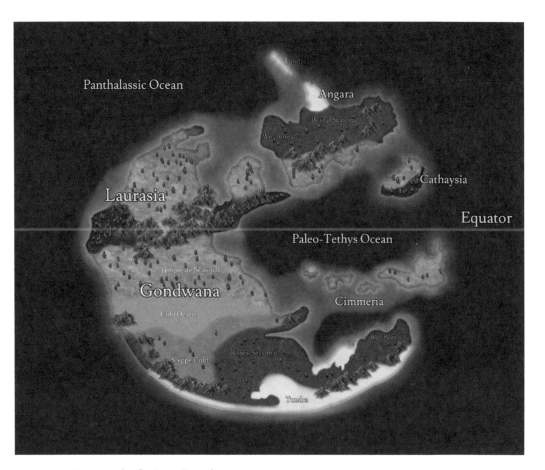

*Pangaea in the Late Permian.*

# THE SCATTERED PIECES OF OLD GONDWANA

While rocks that hold *Glossopteris* fossils are common throughout the scattered pieces of old Gondwana, deposits where plants, insects and other animals are preserved together, and where whole ecosystems can therefore be reconstructed, are extremely rare.

Nonetheless there are some places where glimpses into ancient Gondwana ecosystems can be gleaned. In Morocco, a 2-kilometre thick (1.2-mile) chunk of sedimentary rock called the Ikakern Formation records a snapshot of life in the wet, equatorial regions, with waterways lined with conifers swarming with amphibians and captorhinids. In Niger, which lay further to the south, the remains of a floodplain have captured an arid environment in which pareiasaurs and many small and large therapsids lived. Seasonal monsoon storms swept rain in from the Panthalassic Ocean, bringing life to the dry land for several months a year.

*Glossopteris* remains have not been found in these two northern Gondwana locations, suggesting the forest did not extend north of about 45°S. Further south in Gondwana, however, the Ruhuhu and Usili Formations of Tanzania preserve the remains of dominant *Glossopteris* forests at a paleolatitude of 55°S. Dicynodonts dominated the fauna here, sharing a warm, dry environment with the gorgonopsids that preyed upon them.

Sites in Malawi and Zambia have recorded heavy *Glossopteris* cover in cool, temperate ecosystems dominated by therapsids. Pareiosaurs, which were very common in the equatorial regions, are rarer in these more southerly sites.

In southern Brazil's Paraná Basin, a river delta emptied into a Permian lake. The land here was extremely arid, perhaps even a desert, with a long dry season. Vegetation was sparse but dominated by glossopterids, which coexisted with ferns, sphenophytes and lycophytes.

India in the Late Permian sat at around 55–60 degrees latitude. Fossil sites in the Pranhita-Godavari Valley record life there as being highly seasonal and temperate. *Glossopteris* forests, full of therapsids and captorhinids, thrived in the cool, moist climate.

Eastern Australia holds the richest *Glossopteris* record, and as it lay along the shores of the Tethys Ocean this area likely received large quantities of rain as a result of the monsoon climate in these latitudes. Animal fossils, though, are rare in Australia's Permian age deposits.

In Antarctica's Prince Charles mountains, *Glossopteris*-dominated lagerstätten, laid down in a river floodplain, preserve not only leaves and woody parts of the plants but also the hyphae and spores of fungi that decomposed the forest litter. Insect exoskeletons are also preserved, as well as evidence of their feeding on *Glossopteris* plants.[13]

In a general sense, life on Permian Gondwana was more diverse and prolific the closer to the equator you were. This was a warmer world than the one we live in today, but while mid-latitude areas were crawling with animal life, the forests of the Antarctic appear to have been sparsely populated. No Permian vertebrate fossils have yet been recovered from Antarctic rocks. While it is thought some of the therapsids were warm-blooded, it seems they may have been still too connected to their reptilian ancestry to be able to make much of a life for themselves in these colder polar regions.

When it comes to animal remains, South Africa's Karoo Basin reveals the richest treasures. Here, a vast sedimentary depression that literally covers half of the country records in great detail the lives of the astonishing animals that roamed the *Glossopteris* woodlands of

Gondwana at that time. Ever since Andrew Geddes Bain assembled his treasure trove of fossils, paleontologists have taken countless skeletons from the rocks of the Karoo, devising a system of categorization that now splits the Beaufort sequence into eight different 'zones'.

At the base is the Mid Permian *Eodicynodon* Assemblage Zone, named for the herbivorous creature whose bones are commonly found in this layer. The sequence continues through seven more zones, each named for the animal that is most common in that layer. The final Permian stage of the Beaufort sequence is called the *Daptocephalus* Assemblage Zone, after a dicynodont frequently found in it. *Daptocephalus* lived as one part of a thriving, complex ecosystem that had taken tens of millions of years to evolve.[14] But all of this was hanging in the balance.

The rocks that overlie the *Daptocephalus* Assemblage Zone are very different from those that precede them. In these layers, the first of the Triassic period, the vast array of therapsids that roamed Gondwana are gone. The fossil record reveals an empty, desolate world dominated by just one animal — a dicynodont called *Lystrosaurus*.

And in these rocks, and rocks of a similar age all around Gondwana, the great *Glossopteris* forests, which had thrived for 50 million years, suddenly vanish.

The earliest Triassic deposits speak of a very different time to that which preceded it. They speak of an enormous and geologically sudden change in the world's flora and fauna.

In the final days of the Paleozoic, the clouds of extinction were once again gathering. They would leave their shadow burned into the Earth's crust forever.

# 252 MILLION YEARS AGO

Along the Permian Gondwanide coast, in the region that will one day split off to form New Zealand, a subducting plate drives deep under Gondwana, scraping layers of sedimentary rock up against its side like fat from a fish slice.[15]

Periodically, the volcanic cones of the Gondwanides explode in flames of fire and ash that blanket the surrounding land, choking the forest that spreads away on all sides. At the foot of these mountains, a river descends through cool forests to the coast, where *Glossopteris* trees ring the shores of a cool estuary.

When the tide is full, big amphibians slither along the edge of the harbour, hunting fish and smaller reptiles that throng among the shallows. It is late in the growing season and the *Glossopteris* trees are beginning to shed their leaves, the autumnal foliage rafting along the water and into the estuary. Some of this plant material is buried, becoming part of the slowly accumulating bulk of sedimentary rock forming on the ocean floor. Down here, deep in the Earth's crust, there are slow rumblings, felt only by the fossils that lie buried — the slow, portentous shifting of the interior of the Earth.

For millions of years, a mantle plume, an enormous mushroom-shaped convection cell, has been rising upwards towards the base of the crust. Where it strikes the lithosphere, under eastern Gondwana, it bubbles up the crust and the Earth splits open, liquid basalt, fiery orange, quickly cooling to a hard black rock. Pushed upwards, the supercontinent begins to fracture above the plume's rising fist. Pangaea's time is coming to an end.[16] Further north, another, much bigger mantle plume hits the crust beneath the Siberian Plate. The plume ruptures the surface and enormous quantities of lava spew out, building up a thick layer on the surface.

Gases billow out of volcanic vents in the ground, clouds of it circling the globe, blanketing the atmosphere. Far to the south, on the edge of Gondwana, a coastal *Glossopteris* forest experiences the warmest winter in millions of years. In response to the unusually high temperatures, the trees open their stomata wide, seeking to offload water to cool their tissues. But then spring rains fail to arrive, the river spilling out of the Gondwanides runs desperately low, and the roots of the trees dry out.[17]

Drought creeps across the land. Now the trees go into survival mode. They close their stomata and shut down their metabolic processes. As far as the eye can see a dry, sick desperation overtakes the canopy. The great forests of Gondwana are dying.

# 8

# THE DYING OF THE LIGHT

The story of life on land starts and ends with plants. That is why I chose one to guide us through this journey to Gondwana. The stories of Gondwana and that of the plants and trees that flourished upon it are inextricable.

Gondwana in the Permian was not an isolated landmass. It was connected to the northern continents as part of Pangaea. What defined Permian Gondwana then, was not a rocky shore but the unique *Glossopteris*-dominated 'Gondwanan' biomes that cradled life on the southern part of the Pangaea supercontinent, and which were unique to Gondwana; isolated from their northern counterparts by the belt of aridity that girdled central Pangaea.[1]

Trees shaped Gondwana and were shaped by it in turn. Just as they do today, they moulded the atmosphere of the Permian Earth and in doing so directed the flow of evolution across Gondwana and the world.

Forests are the lungs of the planet. They haul in vast amounts of carbon dioxide and transpire the oxygen we animals need to breathe. Throughout history they have driven the climate as it swung through cold and warm periods, with sometimes violent rapidity. This constant change presents endless challenges and opportunities for the plants and animals that live on our planet. Nothing is ever fixed. Life flows like water across the face of the planet, responding in evolutionary ways to changes in the atmosphere and the movements and capitulations of the ground beneath it.

As landmasses rupture, break, crumple and fold, species are split and separated, and from these fractured lineages new species evolve. New ecological opportunities open up and life moves to fill them. New species arise constantly. And others, many others, disappear from the record forever.

Life, in a broad sense, is incredibly resilient and adaptable, but at the level of individual species, plants and animals are highly vulnerable to change. And nothing, the fossil record teaches us, is forever.

Extinction is a thing so terrible it was once considered impossible, for what loving god would allow it upon the face of their creation? And yet we now know that extinction is part of life on Earth. The fossil record speaks of extinction as a thing so common as to be the form rather than the rule. It is claimed that 99.9 per cent of the creatures that have ever lived on Earth have gone extinct. That extinction rate, however, is not constant. At certain points along life's journey extinction rates have spiked. We interpret these moments as terrible upheavals in the course of life — times when life on Earth was so severely disrupted that huge swathes of species disappeared in a relatively short period of time. We call these events mass extinctions. They appear in the fossil record as sudden absences — rock sequences in which layers crammed with fossils are overlain by rocks that are dead and empty. Periods of time in which not only species, but entire families of species disappear from the face of the Earth, never to be seen again.

These events define the geological timescale. Their occurrences in the fossil record mark the end and beginning of geological periods and epochs. There have been numerous mass extinctions in the history of the planet, but there are five that really stand out from the rest. These 'Big Five' extinctions occur at, and define, the end of the Ordovician, Devonian, Permian, Triassic and Cretaceous periods.

The ocean records these violent death events in the best detail, laying down, as it does, its layers of life-jammed stone, capturing in sediment trillions of swiftly lived lives in the form of empty shells, clean-picked skeletons or deserted coral reefs. And so in the sea is where we can best gauge the impact of mass extinctions. Unpicking the nature of these upheavals on the land is more difficult, as the fossil record is much patchier, but it is clear many of these events have had a huge impact on terrestrial life as well.

Understanding what caused these big extinction events was a key preoccupation of paleontologists throughout the 20th century. Most scientists still veered away from Cuvier's old ideas of massive, sudden 'catastrophes', dismissing such ideas, with their apocalyptic overtones, as alarmist, extremist and perhaps even pseudo-religious.

Slow, gradual processes of extinction and change were, to your average 20th century paleontologist, much more ... seemly. Then, in 1980 a discovery was made that would split the old debate wide open and bring Cuvier's catastrophic outlook to the fore once again.

Mass extinctions, it turned out, did not need to be slow and gradual at all.

# THE SMOKING GUN

For decades, scientists had argued over what caused the dinosaurs to disappear around 66 million years ago, at the boundary of the Cretaceous period and the Paleogene period. Common theories ranged from climate change to greed (they ate themselves out of existence) and even 'racial senility'.

Father and son Luis and Walter Alvarez and their team at Berkeley University wanted to know how fast the dinosaurs' world ended. To do so, they went looking at sedimentary rocks laid down during this crucial time in Earth's history.[2] Sedimentary rocks such as mudstone and sandstone are laid down at differing rates. While 10 centimetres (4 inches) of fine-grained mudstone can be the result of thousands of years of slow deposition on a sea or lake bed, a metre (1 yard) of sandstone could have been deposited by an undersea landslide in just a few seconds. These differing rates of deposition make it difficult to decipher how much time is represented by each layer of sedimentary rock. The Alvarezes proposed that it would be possible to measure the rate of deposition in these sediments by analyzing the amount of iridium in them.[3]

Iridium is a mineral that at the surface of our planet is derived only from extraterrestrial sources — meteorites. Small meteorites hit Earth all the time and leave their signature in the rock in the form of iridium. The Alvarezes' theory was that if you knew the continuous rate of iridium arriving from outer space and measured how much of it was in a layer of sediment, you could calculate how long it had taken that layer of rock to be deposited.

The Alvarez team took samples from a rock sequence in Italy that spanned a huge swathe of geological time, and which included the Creatacous–Paleogene boundary, 66 million years ago. When the team sampled this point in the record they made a surprising discovery. Iridium levels at the boundary soared to 160 times the normal background rate. There were two explanations: either this layer of rock had been lain down over an exceedingly long period of time, or there had been a massive injection of iridium into Earth's geology at this moment. They checked samples from other sites around the world and found the same result — at the exact moment the dinosaurs vanished, a global 'iridium spike' occurred. There was only one good explanation. An asteroid, now thought to be about the size of Manhattan Island, had smashed into the side of the Earth.

Around the same time, a huge crater was discovered in the Caribbean Sea off the coast of the Yucatán Peninsula. A decade later, this crater would be identified as the Cretaceous impact site. The dust and ash created by this astronomical thud would have filled the atmosphere, likely blocking out the sun for many years. Photosynthesis in plants shut down and life on land began to die. The most vulnerable were the biggest animals; the bulk of the dinosaurs were done for. The impact brought the Cretaceous period to its shattered end.

The Alvarez paper shook science to its foundation. Cuvier's much-maligned catastrophism was back, this time with hard geological evidence. Armed with this understanding, scientists could now revisit the other Big Five, looking for similar causes.

## NEVER TO BE SEEN AGAIN

The most sickening dip in the roller-coaster history of life on Earth occurred 252 million years ago at the end of the Permian Period. It delineates the Permian world of the lumbering therapsids, the forests of *Glossopteris* and the great supercontinent Pangaea from the broken, barren days of the early Triassic that followed it.

In the geological timescale, this boundary is so dramatic and important that it separates the Paleozoic (paleo = old) world and the Mesozoic (meso = middle) eras of Earth's history. It's known as the Permian Mass Extinction, or sometimes, 'The Great Dying'.[4]

It is thought that 90 per cent of all life in the Permian Oceans, and 70 per cent of the animals that lived on land, disappeared at this point. It is therefore the largest extinction that has ever occurred on the face of the planet, and we still cannot be entirely sure what caused it. In today's rock record it looks like this: in the sea almost everything is hit hard. Marine rocks from the Late Permian preserve the fossils of thriving reef systems jammed with fish, crinoids, ammonites, trilobites and brachiopods. After the event, the first sediments laid down in Triassic are dark, empty shales — post-apocalyptic mudstones that speak of an ocean starved of oxygen and denuded of life. Whole groups, including the trilobites, sea scorpions and the tabulate and rugose corals disappeared. Other groups, such as the brachiopods, which had been so plentiful in Permian seas, became greatly diminished, huddling at the edges of the ecological scene, just as they do today, 250 million years later.

On land, the fossil record tells a similar tale of woe. The vast diversity of land animals that had evolved throughout the Permian abruptly terminates at the Permian–Triassic boundary. In the younger layers of the early Triassic, most of the Permian synapsids and sauropsids have vanished. Only *Lystrosaurus* thrives.

*Lystroaurus* is a dumpy, odd-looking dicynodont that would seem supremely weird if we were to encounter one today, but in the overwhelming weirdness of the Late Permian and early Triassic would not have seemed anything special. After the Permian–Triassic boundary, in a kind of biogeographical coup that has not been seen

*Lystrosaurus.*

since, *Lystrosaurus*, which was thought to have been a burrowing animal, became exceedingly common. Imagine the strangeness of a world where up to 95 per cent of land animals were, say, sheep. That was the first few hundred thousand years of the Early Permian. There is little to suggest why this genus, among the thousands of others that inhabited the supercontinent in the Late Permian, should have inherited the Earth. But that's exactly what it did.[5]

Maybe it was down to adaptations, such as the ability to burrow away from the hostile heat of a post-apocalyptic sun, or its ability to hibernate, that predisposed *Lystrosaurus* to thrive in these new conditions, or perhaps it was just blind luck. What is certain, however, is that the animals of the early Triassic moved in a world that was severely altered from that which preceded it.

Geological evidence points to a Gondwanan landscape that had become increasingly dry and arid. In sedimentary rocks, the deposits change from organic-rich silts and muds to desolate sandstones. As vegetation died off, there was nothing left to hold the rivers to their meandering streams so they ran wild across the landscape, scouring out huge amounts of material and leaving lenses of gravel to lie upon the land.[6] With the disappearance of forests, the ecosystems that relied on them collapsed.

Three decades after Walter and Luis Alvarez nailed the culprit at the Cretaceous–Tertiary boundary, the search for the cause of the Permian mass extinction goes on. Since the discovery of the Gulf of Mexico crater, many have gone looking for evidence of a similar event, seeking to end the Permian with a bang. But despite the discovery of one or two craters of appropriate age, evidence for an asteroid impact to end the Permian is weak. The theory favoured by most Permian scientists points to plate tectonics. The restlessness of the Earth's crust, which for hundreds of millions of years had driven evolution and fostered life on Earth, now, it seemed, destroyed it.

A huge swathe of eastern Russia is underlain by basalt, volcanic rock that was spewed from the Earth in the form of fluid lava. These lava deposits form an area the size of Australia, around 7 million square kilometres (2,703,000 square miles).[7] This geological feature is called the Siberian Traps and is what's known as a Large Igneous Province, or LIP. LIPs are thought to occur as a result of mantle plumes under the continental crust.[8] As mobile rock expands and is depressurized under a continent, it turns to liquid, breaking through the crust and spilling out across the land like an oozing wound. The Siberian Traps is the largest LIP on Earth. The Traps' eruption, thought to have occurred over a period of two million years,[9] had an enormous effect on Earth's atmosphere. Carbon dioxide, stored for millions of years in the Earth's mantle, was pumped back into the atmosphere in enormous quantities, along with huge amounts of sulphur dioxide and chlorine. The resulting spike in carbon dioxide, along with the other chemicals, warmed the atmosphere and created acid rain, which scoured the land, not only killing plants but also contributing to accelerated chemical weathering of rock. This increased mineral flow into the ocean, leading to algal blooms and a deficiency of oxygen in the ocean.[10]

At the same time, feeder dikes from the Traps — extrusions of lava that spread out laterally from the main eruptive vents — intruded into old coal fields and hydrocarbon deposits underlying the Siberian tundra, releasing masses of the powerful greenhouse gas methane.[11] The heating of the seas may have also unlocked frozen methane deposits along the edges of the continental shelves, which 'burped' to the surface, allowing more of this potent greenhouse gas to escape into the atmosphere, accelerating global warming yet again.[12] The newly methane- and carbon dioxide-rich atmosphere formed a thermally insulating blanket across the planet, and Earth entered a period of rapid global warming.

One devastating effect of this warming may have been a severe disruption of ocean circulation systems. In modern oceans, currents draw warm water north and south from tropical regions towards the poles, where it cools and sinks before it returns along the bottom of the ocean to re-emerge at the tropics as upwellings, rich in nutrients. This aqueous conveyor belt keeps the ocean oxygenated, mixed and healthy. In a warming climate, this temperature-driven system may have shut down, causing the oceans to become stratified, with layers of colder water at depth and warm layers sitting above, and little interchange between them.

Organisms living in that locked-up deep water would then have used up all the oxygen there and died off as those waters became anoxic. Now, the only organisms that could survive in these parts of the sea were bacteria that produce, as a by-product of their respiration, the toxic gas hydrogen sulphide. Over thousands of years, these toxic, anoxic layers grew like a cancer until they filled the ocean and erupted hydrogen sulphide at the surface, pumping even more toxic gas into the atmosphere. The ocean rotted.[13]

In a warming climate, sea water is able to hold less oxygen. In addition, the metabolic rate of sea creatures would have increased. It is possible that demand for oxygen exceeded supply and a huge swathe of marine creatures simply asphixiated.[14] Elevated levels of carbon dioxide would have caused the oceans to become acidic, and creatures whose shells were made of calcium carbonate were no longer able to survive.[15]

The disruption to life on land was immense. There is evidence in the form of mutated pollen grains that the volcanic outpourings might have destroyed the Earth's protective ozone layer, allowing deadly ultraviolet light to rip through the cells of animals and plants on land.[16,17]

Charcoal in the fossil record tells of massive wildfires at the boundary.[18] Spores known as *Reduviasporonites* are found in enormous quantities at the Permian boundary. It is thought these could be the remains of a fungi banquet provided by a feast of dead wood and vegetation.

A huge gap in coal deposition also occurs at the Permian–Triassic boundary, evidence that there were no longer forests, or wetlands to preserve them.[19] For a period after the extinction event it seems most of the world's forests disappeared and were replaced by low, rapidly growing ground cover of herbaceous plants including horsetails, ferns and lycopods. In much of the world, a species of lycopod called *Pleuromia* flourished. This, like *Lystrosaurus*, is sometimes considered a 'disaster taxa', a species that for some reason is able to thrive when everything else is dead.

In the Northern Hemisphere, there was a transition from conifer forests to a flora dominated by lycopods. In the Southern Hemisphere the hero of our story, *Glossopteris*, exits stage left. In almost all the parts of Gondwana where these trees once flourished, they disappear from the fossil record at the Permian–Triassic boundary, 251,902,000 years ago, never to be seen again.

The Permian extinction is thought to have occurred over a period of 30,000 years, a geological instant.[20]

Scientists have concluded that another major extinction event, known as the end-Guadalupan extinction, occurred 7 million years before the end-Permian catastrophe. This event was probably as severe as the extinction event that wiped out the dinosaurs, and certainly had a huge impact on plant life.

How and why the glossopterids should survive this mass disruption, but not the final blow at the end of the Permian is a question paleobotanists have yet to answer. And why should *Glossopteris* vanish at the end of the Permian, while the other groups of plants, such as ginkgos and conifers, survive? Perhaps the answer is simple: *Glossopteris* was a mire species, a tree adapted to long, lazy rivers and cool, waterlogged soil. In the harsh, dry new world these things were in short supply.

A recent study by Chris Fielding and others looked at a *Glossopteris* sequence from Australia and concluded that the plants' die-off occurred 370,000 years before the main marine extinction at the Permian–Triassic boundary. The study found that the extinction was caused by a short-lived rise in seasonal temperatures across eastern Australia. There was no evidence for drastic changes to river action or aridification in their study area. This was not a sweeping apocalypse but a quiet, if rapid, change in Australia's climate. *Glossopteris*, it seems, didn't go out in a fiery bang. Instead, it slipped quietly from the face of the Earth as the climate changed faster than it could handle.[21]

After dominating the Southern Hemisphere for 50 million years, it seems astonishing that *Glossopteris* should disappear with such little fanfare. For the purposes of our story, a mighty asteroid would seem more appropriate. But based on current evidence, that was not the case.

*Glossopteris* simply slips away. And with its passing the old world of the Paleozoic vanished too. A new world was imminent. And in it, our own ancestors would begin to take shape.

# 250 MILLION YEARS AGO
# TO PRESENT DAY

For millions of years the world struggles to find equilibrium. Carbon dioxide levels in the atmosphere continue to fluctuate wildly, hindering the return of forests.[22] The survivors cling to existence in a post-apocalyptic world, dominated by the wandering herds of *Lystrosaurus.*

And then, at last, in the Mid Triassic, around 240 million years ago, the climate settles down and life returns with a vengeance. On Gondwana, the *Glossopteris* fauna is replaced, first by needle-leafed conifers called *Voltziopsis* and later by a flora dominated by another seed fern, *Dicroidium.*[23] In the Northern Hemisphere modern groups of trees, particularly conifers, are now positioned to take over the world.

The age of the therapsids — the group that included the cynodonts, dicynodonts and gorgonopsids — is over, for now. The animals that rise to prominence in these new forests are sauropsids, reptilian descendents of animals that crawled out of those Devonian seas to colonize the land.[24]

The seas too, recover, but with a vastly changed fauna. Modern forms of bivalve and fish now rise to prominence.[25] Gondwana itself is falling apart. Two hundred million years before today it splits from the northern half of Pangaea, Laurasia, bringing the existence of the great continent of Pangaea to an end. Then it splits down the middle, breaking into two chunks: western Gondwana, comprising Africa and South America, and an eastern half comprising the landmasses that will one day be Australia, Antarctica, New Zealand and India.

Between the shore of western India and Africa a long seaway forms, the progenitor of today's Indian Ocean. Forty million years after this great split a new fracture appears, carving off South America from Africa. The waters of the Panthalassic Ocean creep in to form a long channel between the newly separated continents. For a geological while, you might swim from one to the other, then South America vanishes from sight across the curve of the horizon and the Atlantic Ocean is born.

Meanwhile, India breaks to the north and begins sliding across the equator, pulling apart the sea floor that opens up to form the Indian Ocean.[26]

Plants shift and change with the jostling landmasses. From the polar regions to the tropics, the land is heavy with conifers.[27] Cycads, gingkos and horsetails also crowd this warm, wet world. At the equator, forests dominated by *Aracauria*-like conifers tower above swamps and rivers, and provide shelter for the biggest animals that have ever walked the Earth.[28] Sauropsids called dinosaurs have risen to dominance and evolved to fill all the old niches once held by the therapsids. They have expanded across Gondwana, even migrating to polar regions of the Antarctic, ice-free in this sweltering age.[29,30]

A new form of plant has now risen to dominance: the angiosperm. Their key innovation is the flower, a new technology that will define the course of plant evolution from here on in. With this adaptation, plants have found a way to co-opt the help of insects and other animals to spread their pollen. It gives them a massive advantage, particularly in the tropical regions where insects are prolific. Tropical forests that were once entirely dominated by conifers are now constructed of many species of flowering trees.[31]

The asteroid that ends the Cretaceous rips into the side of the planet, wrapping the Earth's wandering orb in a thick pall of dust and debris. When the smoke clears the dinosaurs are gone, along with half the species of plants on the planet. In the aftermath, the angiosperms continue to thrive, outcompeting gymnosperms to dominate the world's flora. Therapsid survivors who trace their lineage back to those old days of the Permian evolve to fill the vacant ecology left by the dinosaurs. Among them, early primates, our tree-dwelling ancestors, emerge in the angiosperm forests of the early Paleogene.[32]

New plant families ripple across the fragmenting Gondwana: *Myrtaceae*, with their fiery blooms and wide, flat leaves;[33] *Proteaceae*, a band of desert warriors adapted to withstand the arid hearts of Australia and Africa;[34] and *Winteraceae*,[35] with their fragrant, spiced leaves.

India's long voyage comes to end as it piles into the side of Asia, thrusting up the Himalaya.[36] The elevation of land that forms the Tibetan plateau profoundly alters weather systems, initializing a monsoonal climate that brings moisture sweeping in off the Indian Ocean.[37]

In South America, the opening of the Atlantic likewise brings huge amounts of moisture sweeping onto the land, allowing the Amazon rainforest to spread across the midriff of the continent.[38] The opening up of the Drake Passage at the tip of South America allows a cold current to wrap itself around Antarctica, sealing in the icy continent.[39]

Conifers and their allies are pushed back to the colder regions, where animal pollinators are scarcer and the conifers' toughness is still an advantage. Bolstered by a new kind of fungal relationship that allows them to draw organic nitrogen directly from the soil, conifers and beeches advance, coming to dominate huge swathes of the Northern Hemisphere.[40] In North America they evolve to form the most enormous living organisms on the planet, the huge redwood trees of the Pacific coast.

In the Southern Hemisphere, the landmasses that were once part of Gondwana are clothed in new forests. In the cool temperate hills of the mid latitudes, on landmasses that will one day form Antarctica, South America and New Zealand, *Nothofagus* forest blankets the hillsides.[41] At the same time an ancient race of conifers unique to the Southern Hemisphere — podocarps, big trees with their ancestry deep in the Triassic — also come to proliferate.[42]

A new kind of plant evolves, one adapted to open country, with a hard brittle structure and a tolerance for cold conditions. These newly evolved grasses spread across the continents, creating wide steppes upon which enormous herds of mammal herbivores migrate and graze.[43]

Our ancestors, the descendants of those early therapsid primates, emerge onto these grasslands, using their fearsome intelligence to hunt the animals that live there.[44]

The physical gap between New Zealand and Australia widens. The land sinks, then reemerges to become cloaked in forest. The continental plate that New Zealand rides upon is thrust up and on top of its neighbour. Mountains climb up on its back, pushed skyward by the contorting plates. As they do, the sky rages against these mountains, breaking them down rock by rock.[45]

In one corner of the South Island, Permian limestone created in an ancient Gondwanan estuary ringed by volcanoes is once again revealed to the sky. These *Glossopteris*-stained fragments of old Gondwana emerge once more at the surface.

# 9
# LINES OF DESCENT

Productus Creek tumbles out of the Takitimu Mountains in southern New Zealand, little more than a trickle against the bulk of these hills. It carves its way down a steep hillside cloaked in a drab, grey matagouri scrub, invasive briar and dripping native tussock.

*Nothofagus* forest crowds the valley's opposite side. The land feels old and cold, silent, desolate with absence, as if aching for the return of its long-vanquished birds — the moa, Haast's eagle, and the adzebill.

The Takitimu Range was named by the first Māori who arrived here. To them it was the upturned waka, or canoe, of their ancestors. Like all of New Zealand's mountains, the range has been thrust skyward by the torturous geological contusions that keep our country aloft.

New Zealand is part of a continent called Te Riu-a-Māui/Zealandia, which if it were all above water would be about the size of Australia. Most of this continent, however, lies submerged beneath the Pacific Ocean and the Tasman Sea. The bit that sticks out above the water — what we now call New Zealand — straddles the boundary of two clashing tectonic plates. It is these rigid forces, quivering with stored potential energy, that hold us skyward in our fragile poise.[1]

We live in fear of the next shifting of the ground, of the Big One; the moment the Alpine fault, which runs the length of the South Island, makes one of its periodic adjustments, flattening cities, destroying roads, killing us. New Zealand's history is punctured by earthquakes. A recent earthquake in 2011 destroyed much of the city of Christchurch and took 185 lives. In terms of earthquakes, this was a relatively minor wobble.

The bulk of the rocks that make up New Zealand were laid down off the coast of Gondwana between the Permian and the Cretaceous as marine sediments. They are greywackes — heaps of sand and mud deposited by enormous undersea landslides called turbidite flows. Turned to stone over time and thrust skyward by the mountain building 5 million years ago, they now crumble away, eroding from the sides of these mountains in riverbeds that grumble with the weight of the gravel and boulders they constantly move.

At the edge of these terranes, older volcanic rocks, like those of the Takitimu Mountains, are present. These rocks were spewed out by volcanoes along the margin of old Gondwana. They are evidence of old subduction zones, of turbulent geological life at the edge of Gondwana.[2]

Nestled in the nook of these volcanic hills is the Mangawera Formation, a limestone laid down as shallow sediment in an estuary that drained off the back of these old volcanoes in the Permian.[3]

I've come here with geologists Hamish Campbell and Chuck Landis. For Chuck, this is a return to old stomping grounds; the retired geology professor used to bring his students up here to get a feel for one of New Zealand's most important fossil localities.

*Chuck Landis (left) and Hamish Campbell at Productus Creek.*

Productus Creek is such an insignificant trickle that it might not even warrant the status of a name were it not for the geological importance of what its persistent grinding has revealed. The stream cuts into and exposes marine deposits clogged with seashells, mostly those of the ancient brachiopod *Productus*, which lends its name to the creek.

Because New Zealand was little more than a wedge of debris pushed up against the side of Gondwana in the Permian, this country has few known Permian terrestrial deposits; almost all of the country's sedimentary Permian rocks were formed on the seabed. What makes the Mangawera Formation particularly interesting is that it wasn't laid down in the deep sea but in a shallow estuary near to shore, and because these sediments were put down so close to land they contain within them rare fragments of the plant life that lived on that coast, including, very occasionally, *Glossopteris* leaves.

Only a few have ever been found here — just enough for New Zealand to be officially part of the *Glossopteris* club along with the other ex-Gondwana countries. It was from here that the *Glossopteris* specimen that started this whole journey for me, the one displayed in the University of Otago's geology museum, was recovered.

We've come here looking for more. I follow Chuck through the dense scrub, fighting a tangle of thorns and pushing our way through big, sopping tussocks, looking for exposed rock outcrops among the dense vegetation. Finally Chuck locates a suitable stub of rock sticking out between a couple of big matagouri bushes and gets hammering. His blows reverberate across the valley, filling the mountains with alien sound. The rock is tough, dark and gnarly, hard to split. After persistent smashing, it breaks off in jagged arrows that drop to the muddy ground at our feet. Chuck picks one up and passes it to me.

'Smell it,' he says.

I put the rock to my nose and inhale. It smells like a mechanic's concrete floor, cold with winter, sharp with petroleum. 'Hydrocarbons,' Chuck explains. So much living material was being washed off the land that time and pressure have preserved its smell in the rock. What I can smell is life — the organic essence of that vanished supercontinent, preserved for 250 million years, chemically altered and turned to stone. Living rock. Old, buried carbon. What we seek when we tunnel into the Earth in search of fuel to power our machines. Not that these rocks will power much — their carbon content is far too low to be of any economic value.

We keep bashing away at the rock. I find plenty of brachiopods but nothing that springs out to me as plant material. Suddenly, Hamish calls across to us. I go over to look at what he's found. He has split open a chunk of rock and there, rippling across the split is a dark shadow that, if you hold it to the cloudy sky in just the right way, looks a lot like a leaf.

I've seen enough *Glossopteris* fossils by now to recognize the venation pattern, the elongated curve of its edge. So here it is, the mythical leaf I travelled halfway around the world to find, broken out of Permian rock in my own backyard.

This is not like the perfectly preserved *Glossopteris* leaves I saw in Argentina or the Falkland Islands. It's a fragment, a watery whisper of the past, just enough to remind us that Gondwana may be gone but we retain its ecological and geological heritage. I run my fingers over the rough stone, imagining it as soft mud, part of a living world that has now vanished.

I set out to find Gondwana, imagining I could go there; understand its shape, its climates, its moods. In looking for Gondwana I have travelled halfway around the world, trawled through hundreds of scientific papers and books and talked to dozens of scientists. It has at times been a frustrating experience. The scientific literature is dense and prolific. After hours of reading I was often left with my head drowning in detail.

And then there are the differing points of view. Every time I thought I understood something about what happened on Gondwana I'd stumble on a paper that claimed to have shown the opposite.

Absorbing everything I'd read, I tried to travel to Gondwana in my imagination, applying my meagre knowledge of science, ecology and the natural forces of the modern world to a place that hasn't existed for many millions of years.

And yet Gondwana still remains frustratingly out of reach. She is everywhere and nowhere. Through the fog of understanding I can see her vague shape. I can comprehend her in an academic way. But I can't get there.

But then, maybe that's the nature of paleontology. All across the world, scientists go in search of these vanished ancient places. Like me, they get lost in detail as they catalogue, describe and list. I wonder if any of them really get close to truly understanding the places they seek.

I realize the best I can do to understand Gondwana is to understand it as it is now. Our fragmented Gondwana. The place I grew up on.

And just as plants have been my tour guide to the vanished Gondwana, it is to them I now turn to understand the modern Gondwana and the ecological forces that sustain her.

# THE GENETIC SIGNATURE OF A LONG JOURNEY

At the latest count there are 391,000 vascular plant species in the world.[4] They rampage across the broken pieces of Gondwana and the rest of the planet too.

In South America, the greatest rainforest on Earth, the Amazon, grades into the driest desert on Earth — Chile's Atacama Desert where, despite its crippling aridity, plants also survive. In Africa, plants occupy the Sahara Desert and also form rainforests in the Congo Basin. In Australia, plants have adapted to aridity and fire, developing the hard leaves and water retention capability that allow *Eucalyptus* species to dominate the red continent today. Thick tropical forests blanket Papua New Guinea.

There are an estimated 3 trillion trees in the world,[5] comprising more than 60,000 species.[6] They each carry within them the genetic signature of the long journey plants have taken, from the earliest algae that established on the rocky seashores of Gondwana hundreds of millions of years ago to the towering behemoths that dominate the world's greatest forests today.

The biggest trees on Earth are the giant redwoods of California, which soar to almost 116 metres (381 feet), the length of a football field. It is possible even larger trees once stood in the forests of North America but were cut down for timber.

In my own country, a kauri tree was once recorded that grew to 65 metres (213 feet) and was the width of a two-

*The world's biggest tree, the General Sherman giant redwood, California.*

lane highway. This tree was 4000 years old.[7] Trees face a conundrum in getting as big as this. On the one hand, their huge size allows them to tower over their neighbours and monopolize the available sunlight; but on the other, being so large makes it incredibly hard for the tree to draw up water through its xylem tissue to the uppermost leaves. The giant Californian redwoods and Douglas-fir trees, which historically attained even greater heights, are thought to approach the maximum size that a tree can physically attain, the point at which it becomes inefficient or impossible to draw up water any further.[8]

As trees grow, they lay down successive layers of xylem that appear in a cross-section of the trunk as rings, and which can be used to date them. The longest-lived tree on Earth is thought to be the bristlecone pine of Oregon, which can attain an age of 5000 years.

Trees have evolved into niches that span much of the globe. At the equator, tropical forests are dominated by angiosperms and characterized by staggering levels of diversity. In the Amazonian rainforest alone there are almost 7000 species of tree.[9] In the higher northern latitudes, sprawling boreal forests are far less diverse, dominated by just one or two species of conifer, such as spruce or fir.[10]

Trees have been around for 300 million years, in essence virtually unchanged. A tree today looks and acts basically the same as a tree that existed in the Permian. They are one of evolution's most successful achievements, a design so effective there was no need for improvement.

Tree by tree, generation by generation, forest by forest, they have ridden the movement of the continents across the face of the Earth.

# FLOATING OR FLYING

After the break-up of Gondwana, New Zealand drifted off on its easterly course, its plants and animals heading off on a unique evolutionary path which produced a flora and fauna unlike anything else on Earth.

When the first humans set foot on the shores of New Zealand, they discovered a land clothed in dense forest and filled with an unimaginable array of birdlife, one of the last true wildernesses left on the planet. Those first people to arrive, the ancestors of today's Māori, walked through a landscape that had been evolving in isolation from the rest of the planet for millions of years. The coastal and lowland areas were dominated by enormous podocarps — kahikatea, matai, miro, rimu — among them trees that could live for thousands of years.[11]

In the northern part of the country, a great kauri forest covered the land. The hillsides of the South Island, prone to heavy snow and frost, were cloaked in a forest largely dominated by podocarps and four species of beech in the genus *Nothofagus.*

There were no mammals, save two species of bat. Every ecological niche that mammals fill elsewhere in the world had been taken up by birds, many of whom, in the absence of ground-based predators, had abandoned the ability to fly.

There were birds that snuffled along the forest floor in search of grubs, in the same way a rat or a vole might elsewhere. There were birds that swung through the trees, using their beaks as hands, in the manner of monkeys. In place of lions or wolves, there were giant carnivorous eagles, the biggest that ever flew, stalking the skies.[12] The forest would have shrieked and yammered with a cacophony of bird sound we cannot possibly imagine today.

Within a few hundred years of human arrival, a third of these forests were gone and many of the species that lived among them were driven to extinction, including the giant moa.

My own ancestors arrived here from Europe around 200 years ago. They came to carve for themselves a new life from what remained of the wilderness. In doing so, they set out to recreate the neat, green, ordered land they had left behind. Implicit in this was the deep fear and hatred of the wilderness that still loomed so hugely in their psyche.

And so they attacked it with saws, axes, and bullocks and chains. In just a century, most of the remaining forest had been destroyed, cut down for the valuable timber it contained and cleared to make way for farmland. The settlers introduced new animals that reminded them of home, like rabbits and deer, which exploded out of control, plundering the natural capital of the land. Predators like cats, rats and stoats came too, and made short work of what was left of the native fauna. New Zealand was left destitute. Today we are learning how to save what we have left. At the same time, we are peering deep into the past, using genetic tools to figure out how our native plants, birds and animals came to be here.

Researchers talk of many of New Zealand's native plants and animals as having 'Gondwanan' origins. What they traditionally meant by this is that these species originally arose on the supercontinent, then when Gondwana fragmented they were carried along for the ride. This process, known as vicariance, explains the distribution of these species and their relatives across the old Gondwana fragments, at least according to established 20th century ways of thinking. *Nothofagus*, for example, is also found in South America, Papua New Guinea and Australia and so its presence in these places has been seen as being an artefact of their former connection. Vicariance has also been used to explain the distribution of families like *Proteaceae*, which includes South Africa's famous proteas and the banksia trees of Australia.

Vicariance theory harks back to Hooker, Suess and Wegener. The latter two leaned on vicariance when they looked at the distribution of *Glossopteris* fossils in widely separated landmasses and theorized that this was evidence these landmasses had once been connected. For 180 years, vicariance was the preferred explanation for such distribution. But there is a problem with it.

Vicariance presupposes that most plants and animals are incapable of crossing large stretches of ocean. To a degree this is logical. *Nothofagus* seeds, for example, cannot survive immersion in salt water. But is that enough to discount the idea that such species can move across it?

Charles Darwin was an early proponent of dispersal — the idea that species were in fact capable of colonizing landmasses on the other side of an ocean by themselves, by floating or flying across the water. And in recent decades, as molecular science has caught up to the debate, the pendulum of opinion has swung away from vicariance and back in favour of dispersal. Scientists are now able to compare genetic material from closely related species of plants and animals and, by using molecular clock techniques, calculate how long ago these species separated from each other. What they are finding is turning the old ideas about vicariance and 'Gondwanan distribution' on their head.

A recent study by Michael Knapp and others looked at the relatedness of *Nothofagus* species across Gondwana and found that *Nothofagus* trees in Australia and New Zealand actually separated far more recently than a Gondwanan split would allow. New Zealand has been separated from Australia for 80 million years and yet the *Nothofagus* trees that the research group studied had only been separated for around 30 million years.[13]

Another recent study looked at the divergence of plants in the *Proteaceae* family, another classic 'Gondwanan' species, and found a similarly late split.[14] Studies now show plant species have regularly dispersed across oceans since the break-up of Gondwana.

To the east of New Zealand lie the Chatham Islands, volcanic blisters that were thrust up out of the ocean just 4 million years ago. And yet, despite their youthfulness, the Chathams have a full complement of plants, insects and reptiles that can only have got there by floating or flying.[15] What all this shows is that gene flow across oceans is not only possible, it happens quite often. How these species cross the ocean is a matter for conjecture. It may seem unlikely that a plant whose seed cannot survive in salt water should be able to cross a sea to colonize a landmass, but that's only because it is difficult for us to imagine the enormous amount of time these plants have had to find a way to do so.

Our minds can conceive of a length of time equivalent to our own life span, that of our parents and perhaps that of our grandparents. But even going back four or five generations, time becomes hazy to us. So how much harder is it to imagine how long a million years is, let alone 30 or 300 million years?

The ability of a plant or animal to cross such a seemingly insurmountable barrier as an ocean relies on chance events, highly unlikely occurrences. In the case of a plant, that might mean a seed gets eaten by a bird, which is then caught up in a big storm that sweeps it across thousands of kilometres of ocean to deposit it in a pile of guano on the shores of a different landmass. It might mean a one-in-a-thousand-year flood that tears a huge island of soil and living vegetation from a riverbank and washes it out to sea, where fortuitous winds and currents carry it for thousands of kilometres. Aboard such a floating raft might be birds or animals, fungi and insects that are able to survive the voyage and colonize the new land where the raft washes up.

Such events are most likely incredibly rare. However, 'incredibly rare' means nothing in the context of geological time. If something has a one-in-a-million chance of happening in any given year, then statistically it might have happened 80 times since New Zealand broke away from Australia. Now, we start to see that 'unlikely' has no resonance in the context of our story. Plants and animals can cross oceans. Even the moa, New Zealand's famous giant flightless ratite, has been shown to have descended from a bird that flew here around 40 million years ago.[16]

Gondwana, then, in the sense of being a physical landmass, may be long gone, but Gondwana in the sense of a biogeographic province is still in existence, as ever since its break-up, animal and plant genes have continued to flow between the fragmented landmasses that were once part of it.

Gondwana lives on.

# BRANCHES ON THE TREE OF LIFE

I enter the podocarp forest of Rakiura/Stewart Island, in southern New Zealand, on a drizzly day. The rain trickles down through the canopy, irrigating a lush tableau of vines, lichens, mosses and shrubs. My feet squelch into earth thick with matter.

The air smells of decomposition and fresh growth in one wet, woody breath. At the base of a big podocarp I drop to my knees and push my hand into a thick mat of growth. This moss is a descendent of some of the first plants to ever colonize Gondwana; 400 million years later, they are still here, still reproducing with spores and water in their ancient way.

Further along the trail, a long tangle of foliate fingers tumbles from a bank — this is a native lycopod, or club moss. These, too, are living relatives of some of the first plants to arrive on Gondwana, and of the vast Carboniferous *Lepidodendron* forests which once filled the tropical world. In the modern world, club mosses are no longer canopy-dominating giants. In fact, they don't get much bigger than a metre (1 yard) or so in length. Once again, they are small beings, content to live in the shadow of the forest giants.

These giants are the southern family of podocarp trees. They tower above me — totara, rimu and matai — big, ancient trees, some of which may well have been alive before humans came to New Zealand. Podocarps stretch back in the fossil record to the early Triassic. Their ancestors were among those that inherited Gondwana after *Glossopteris* had vanished. It's possible they might even be its direct descendants. We may never know for sure. Teasing apart lines of evolution in the fossil record is extremely hard and always contentious. And while it is intriguing to think in terms of direct lines of descent, it is also, to a degree, academic. Evolution is not necessarily a linear progression in which x leads to y leads to z.

In evolutionary terms, *Glossopteris* species are more accurately imagined as twigs at the end of branches on the tree of life. While individual twigs may have ceased growing, the branches they were attached to remain very much alive and continue to put out new twigs all the time. This branch, at its thickest point, connects every living plant species in the forest of the Permian to the forest of today. So, as I walk through this dripping New Zealand rainforest, it is certain that the genetic instructions written into the DNA of those ancient *Glossopteris* trees are still alive in every trunk, branch, leaf, flower and root of this forest. All around me, the same genes that inhabited *Glossopteris* stems on Gondwana are still at work, furiously reproducing themselves down here at the edge of the world.

Gondwana lives. And those long-lost *Glossopteris* trees that flourished along the edges of Gondwana's rivers and lakes in the Late Permian? As it turns out, they're not gone, either.

# PRESENT DAY

Shrouded in darkness, Planet Earth turns beneath us, the shapes of her continents marked out by galaxies of artificial light that create from the darkness the shapes of Africa, the Americas, Arabia, Europe and Asia. Only uninhabited Antarctica remains invisible against the darkness of the nighttime sea.

Now the sun breaks the skin of the Atlantic Ocean, its westering rays sweeping across the long, lonely steppe of Argentina and striking the side of the Andes — the cordillera that forms the snowcapped spine of South America. The day's first light reveals the trailing edge of a continent-sized ship — South America, still reeling from its wrenched separation from the rest of Gondwana. For 50 million years it has been drifting westwards, travelling 5000 kilometres (3100 miles) (relative to Africa) across the face of the Earth, at a pace of around 2 centimetres (0.7 inches) a year, slower than the speed a human fingernail grows at.[17]

On its leading edge, beyond the mountains, the South American continent rides over the oceanic crust of the Nazca Plate, which underlies a portion of the Pacific Ocean. The eastern edge of the Nazca Plate is slowly being drawn down and under South America, to be subsumed back into the mantle from where it came.

Here, volcanoes puncture the ground above the subduction zone, forming the long arch of the Andes that stretches all the way from Tierra del Fuego at the southern tip of Argentina to Ecuador in the north. Occasionally these Andean volcanoes still explode into life, hurling massive amounts of ash out across Patagonia, a thick grey smothering blanket that renders the land sterile for years to come.

Africa, meanwhile, pushes northward on its slow collision course with Europe. All around the Pacific, subduction zones swallow oceanic crust on all sides, and from Japan to Washington State, from Indonesia to New Zealand, volcanoes light up the 'Pacific ring of fire'. In the midst of the Pacific, islands bubble up where mantle hot spots occur, erupting as slow volcanoes in places like Hawaii and the Aleutian Islands.

We look down on a patchwork quilt world, much of the land cleared of forest and shaped by agriculture. Great, seething cities erupt from the land, the air above them thick with pollution. Only in the Arctic do unbroken forests still hold sway. In the Americas, the Amazon rainforest is besieged on all sides by smoke and flame.

All across this planet, 7.8 billion people scrabble for existence.

# 10

# RECKONING

In a dark, cramped 'rat hole' dug into an abandoned opencast coal mine in Jharia in the northeast of India, a child scrapes at the ground with calloused fingers. Into her plastic basket she scrapes a load of crumbling coal before crawling back out of the treacherous hole. She stands, hitching up her ragged dress and walks on bare feet up the steep side of the escarpment. This small heap will earn her a few meagre rupees when she sells it to mafia men back in her village.[1]

The side of the mine pit crawls with people all doing the same. They are risking injury and death, from toxic gases, cave-ins and cliff collapses, to be here. They are also breaking the law, stealing from the big coal corporations who have moved onto richer fields, neglecting to fill in this worked-over pit. These people, among the Earth's most wretched, do what they must to survive. They live in a kind of post-industrial hell — the 21st century incarnation of Suetonius Grant Heatly's coal-mining dream. Their village teeters on the edge of a wasted country. Beneath the very ground their houses stand on, subterranean fires that have raged for over a century eat away at the coal seams. Periodically, holes open up in the Earth, swallowing houses, animals and people.

The air is dense with coal smoke and thick with chemicals. It coats the insides of these peoples' lungs and poisons their lives. They contract tuberculosis, cough blood and die young.[2]

This is life in the belly of India's coal industry. Jharia coalfield, in Jharkhand's Damodar Valley, is India's largest. A century after it was first mined, it is estimated almost 20 billion tons of coal still lie in the ground here.[3] Many of the people who live in the area are employed by the industry in some way, either working for big coal companies or scratching a living from the ruins.

The state-owned Bharat Coking Coal Ltd (BCCL) is responsible for much of the opencast mining in the region. In 2019, BCCL processed $2.2 billion worth of coal. Little of that wealth trickles down to the people of Jharkhand, though; almost half of the region's inhabitants live in dire poverty.[4]

Most of Jharia's coal is low-grade coking variety and used to make the steel needed to structure India's rapid industrialization and urban growth. The city of Jharia and neighbouring Dhanbad are the most polluted towns in India, with particulate matter in the air three times the safe level.[5]

India's vast coal reserves lie in seams beneath the Gangetic Plain, crumpled along with the northern edge of the country into the fold belt of the Himalayas and extending into the south of India. This coal is almost entirely of Permian age.[6] It was formed from the remains of *Glossopteris* forests that once flourished on Gondwana. 70 per cent of India's energy is created by burning this coal, which releases all the carbon captured by those trees and buried in the ground 250 million years ago back into the atmosphere.[7]

India consumes more coal than any other country on Earth, except for China, and is one of the world's top coal polluters.[8] Yet, despite the vast amount of coal India digs from the ground in places like Jharia and neighbouring Raniganj, it still demands more. Much of that coal comes from Australia,[9] which produces more than 400 million tons of bituminous coal annually,[10] most of it from Permian coal deposits along the country's eastern states. Opencast mines in Queensland and New South Wales work those same Gondwanan *Glossopteris* coal seams as are found in India, only now separated from them by the Indian Ocean.

The Bowen Basin, which straddles the border of New South Wales and Queensland, contains Australia's largest coal reserves and is one of the world's largest deposits of bituminous coal.[11] Further south, the Sydney Basin is a swathe of bituminous coal which underlies the country's biggest city and outcrops along the New South Wales coast between Wollongong and Lake Macquarie — the homeland of the Awabakal people, who first knew of the region's coal and wove it into their mythology. Even after two centuries of heavy extraction, the basin is estimated to hold more than 7 billion tons of recoverable coal. It's a resource that pumps around $15 billion of value into the Australian economy each year and which provides 80 per cent of New South Wales' electricity.[12] The coal from the Sydney and Bowen Basins is excavated from opencast pits inland and loaded onto trains that carry it to the coast, where it is loaded onto ships that carry it across the ocean to bolster India, China and Japan's inexhaustible hunger for coal. Meanwhile, Australia still relies on coal for more than half its electricity needs.[13]

In South Africa's Highveld coal belt, where many coal mines pock the land and petrochemical plants and coal-fired power stations are rife, hundreds of people die each year from lung cancer, heart disease and other diseases related to toxic levels of sulphur dioxide and nitrogen dioxide in the air.[14] Almost 90 per cent of South Africa's electricity generation comes from Permian coal.[15,16] Without these ancient *Glossopteris* forests to burn, South Africa might crumble into economic ruin.

*Glossopteris* remains now form most of the Southern Hemisphere's coal reserves, while much of the hemisphere's offshore oil and natural gas is also derived from those ancient forests.

All across the world, coal use has soared in the last 20 years, driven largely by the rapid growth of the developing nations and the massive economic expansion of China and India. Burning coal injects enormous amounts of carbon into the atmosphere, which binds with oxygen to form carbon dioxide, the greenhouse gas most responsible for global warming. Sulphur and nitrogen, which occur as trace elements in coal, combine with water vapour to form acid rain.

Coal mining releases into the atmosphere methane gas trapped in the coal seams — millions of tons of it each year. Methane is one of the most potent greenhouse gases known.

Forty per cent of the emissions from energy use come from burning coal.[17] As a result of the huge amounts of carbon dioxide we have pumped into the atmosphere since the start of the industrial revolution, carbon dioxide levels in the atmosphere have risen dramatically since at least the 1960s.[18] Global temperatures are rising. The planet may be the hottest it's been for 125,000 years.[19] All around the world, glaciers are in rapid retreat[20] and ice caps are melting.[21] Coal use is the single biggest contributor to the planet's current temperature rise.

This might be starting to sound like a familiar scenario. At the close of the Permian, volcanic eruptions spewed carbon dioxide and other greenhouse gases into the air to bring about catastrophic climate change. Now we are doing the same, through artificial means, by burning fossil fuels, and especially coal.

Today, atmospheric carbon dioxide content is 410 parts per million, the highest it's been in 3 million years. While we still have a long way to go to reach Permian extinction levels of greenhouse warming, when carbon dioxide levels soared to more than 2000 parts per million,[22] it's worth remembering that the Permian extinction was just the biggest and most drastic of many climate crises throughout Earth's history.

There are many more mass extinctions recorded in the fossil record. The Ordovician–Silurian extinction. Olson's extinction. The Cenomanian–Turonian boundary event … the list goes on. Almost all of them are associated with climate change and all led to widespread extinction.[23] The message in the rock record is quite clear: life is incredibly vulnerable to climate change.

The Permian extinction was a huge upheaval in the structure of life on Earth, changes that were enough to drive *Glossopteris* — a genus that had thrived for an enormous amount of time — into oblivion. The great irony of our current ecological crisis is that by burning the coalified remains of these tree species, which were destroyed by global climate change, we may be precipitating yet another climate change crisis in our own time. And at the same time as we are burning up these old fossilized forests and filling up the atmosphere with the stored carbon they contain, we are destroying living forests at a frightening rate. In doing so, we destroy our planet's ability to counteract the crisis.

*Coal scavengers, Jharia, India.*

Forests, particularly those at the tropics, are the most important sink of carbon on the planet. The planet loses around a football pitch of primary forest every six seconds.[24] Old-growth forests continue to be logged in North America, while illegal logging is rampant throughout the tropics. In these regions, whole swathes of old-growth rainforest are being cleared to make way for palm plantations.

Wildfires rage in the Arctic and the Amazon, burning peatland and forest. Wildfires have in recent years also scorched huge swathes of Australia and California as atmospheric temperatures rise. The incidence of wildfire in the United States is increasing.[25] When fire rips through a forest it not only destroys an important carbon sink, it also converts the carbon locked up in that forest to carbon dioxide in the atmosphere, further accelerating global warming. Such warming is melting permafrost in the Arctic, releasing frozen methane into the atmosphere,[26] while the reduction of snow and ice cover in the polar regions causes the atmosphere to absorb more of the sun's heat.[27]

The world is changing, fast, and these negative feedback loops are all a part of the global climate change nightmare.

There are some scientists who argue we are in the middle of the sixth great extinction.[28] If *Glossopteris* could be wiped from the face of the Earth after 50 million years of existence, why should we, with our meagre 50,000 years of time here, feel invulnerable?

It's a dire picture, but there is some hope. It seems that trees may be coming to the rescue — of the planet at least.

As carbon dioxide levels rise, a recent study has found that trees are photosynthesizing at a faster rate.[29] And against all odds, the Earth is getting greener. This is reported by studies, which have looked at forest cover across the planet and found that, overall, it is expanding.[30]

Despite the devastating deforestation reported in many parts of the tropics, forests are regaining ground in the temperate regions. This is partly down to changing land use. As our cities become better organized, agricultural production is concentrated in smaller, more productive areas, and less productive areas in hilly and mountainous regions are abandoned to wild forests.

Affluent countries also place more of a focus on rewilding because they can afford to do so, and so are more likely to lock up areas of land from destruction. Then there are commercial forests, mostly of conifer species, which, although low in biodiversity, also form a significant carbon sink.

Trees are on the move. Studies have recorded trees migrating northward into Arctic areas and in the United States westwards, at a rate of around 100 metres (328 feet) a year,[31] and perhaps much faster. Trees, of course, cannot migrate themselves. The movement of a forest requires an intergenerational leap as seedlings establish in areas that are more favourable, and fail to germinate in areas that are not. If the favourable areas happen to be north of the existing tree line, and the unfavourable areas are to the south, the net effect will be a northward movement of the forest. As the world gets warmer, forests are fleeing the heat of the tropics and pushing into more northerly areas.[32]

Is it possible the planet is seeking equilibrium? British scientist James Lovelock is best known for his famous Gaia hypothesis, a theory he presented in the 1970s which sees Earth as a self-regulating system geared towards nurturing life on the planet. In essence, Gaia theory presents Earth as a kind of superorganism, a complex system in which every part operates to tune the planet's climate in favour of the preservation of life. What Lovelock suggested is that there are naturally occurring checks and balances that keep the Earth's atmosphere and biosphere habitable for life. These processes are driven by life itself — not consciously, but simply through the way species interact with their environment. These checks and balances are what maintain our life-giving atmosphere.[33]

Earth is thought to be the only planet in the Solar System with a habitable atmosphere. Venus has a thick, toxic atmosphere thanks to a runaway greenhouse effect that now sees temperatures at its surface soar to 480°C (900°F). Mars is thought to have once had an atmosphere but for reasons unknown it has vanished, leaving the red planet as a desolate, cold wasteland.

Were negative feedback loops to spiral out of control and carbon dioxide levels in Earth's atmosphere to climb to astronomical levels, as seems to have happened on Venus, our planet could become a ball of toxic gas that sustains no life. Similarly, if carbon dioxide levels ever plunged too low, the global temperature could cool so drastically that we might be plunged into a permanent 'Snowball Earth' state.

Yet, despite the wild climate fluctuations throughout Earth's history, this has never happened. Every time there was a massive carbon dioxide spike, such as that which occurred at the end of the Permian period, natural systems have eventually remedied the situation, drawing all that excess carbon dioxide out of the atmosphere and returning the Earth to a more moderate temperature.

Plants, as we have seen, are central to this. Plants, and the soils they create, act as a natural thermostat for the planet. When carbon dioxide builds up, the planet warms and plant life thrives, helping to remove excess carbon dioxide from the atmosphere. When too much carbon dioxide is removed from the atmosphere the planet cools, plant growth is stunted and weathering is reduced. As a result the carbon dioxide drawdown falters and the world begins to warm again.

In 1983, Lovelock, along with Andrew Watson, expanded on the Gaia hypothesis by presenting a scenario they called 'Daisyworld'. Daisyworld is encapsulated in the idea of an imaginary planet upon which grow only two species: black daisies and white daisies. White daisies thrive in warmer conditions and, being white, have high albedo — that is, they reflect a lot of solar energy back into space. Black daisies, on the other hand, thrive in colder conditions and absorb much of the sun's heat. When temperatures on Daisyworld get too cold, black daisies will thrive, covering the world in heat-absorbing ground cover that will preserve the sun's energy and have the net effect of warming the earth. Over time, temperature will return to a more moderate level. If, however, the earth gets too hot, white daisies will proliferate, covering the landscape in reflective white. Rather than being absorbed by the planet's surface, the sun's energy will be bounced back into space, and over time the planet will cool.

Daisyworld is therefore a self-regulating system in which the planet is maintained at a habitable temperature level by the species that inhabit it. Could it be that a Daisyworld scenario is occurring on Earth, and that, just as on Daisyworld, it is plants that control it?

Paleontologist Greg Retallack has proposed just such a thing.[34] By studying soils through Earth's history, he has found that during times of high carbon dioxide and therefore high temperature, tropical forests, and their carbon-hungry soils, spread into the polar regions, eventually drawing in the excess carbon dioxide and cooling the planet. The migration of trees and forests into Arctic regions (and, at the same time, the expansion of grasslands and their soils into desert regions), Retallack believes, will ultimately have the effect of cooling the Earth. In the long run, the northward expansion of forests may help pull Earth back from a Venus-like catastrophe.

How the expansion of tree cover into the Arctic will initially affect global temperature depends on complex dynamics that are poorly understood. Other studies suggest that, at least in the short term, the migration of trees into Arctic regions may in fact lead to a net increase in warming.[35]

We live in a nexus time, when our understanding of Earth's processes is coming together with our understanding of how drastically we have affected them. A thousand scientific journeys, small and great, from Leonardo's seashells and Cuvier and Brongniart's studies of the Paris Basin, William Smith's geological maps, Scott's *Glossopteris* fossils and Wegener and Du Toit's recreation of an ancient supercontinent, have converged on this point in history.

It took hundreds of years to conceive of Gondwana and Pangaea, and the idea that continents could move across the face of the Earth. Hundreds of years to understand that the Earth's life systems are in constant flux, and that extinction is real. It has taken hundreds of years to reach an understanding of the complexity and interconnectedness of Earth's systems — how the rock cycle and the carbon cycle interact and how they drive the evolution of life on Earth. Only in the last 50 years have we arrived at continental drift. We have also arrived at a deeper understanding of forests, and the role they play in the Earth's ecosystems and how crucial they are to the health of the planet.

In the long run, expanding, rapidly photosynthesizing forests may draw down enough of the carbon dioxide we have released into the atmosphere to save the planet from a runaway greenhouse effect, but by that stage, we may well have severely compromised our own ability to survive on this planet. We can't just sit back and wait for trees to save us — trees and forests don't operate on human timescales.

Critical to putting the brakes on climate change is easing our dependence on coal. And here, there is some cause for hope. More affluent countries are already reducing their demand for coal. China, the world's biggest user of coal, has launched an ambitious plan to transition to renewable energies in the next few decades.

Though not without their own issues, renewable technologies like wind and solar are becoming more prevalent and more efficient. Maybe we will find a way through this crisis — or maybe not. But analyzing that is a topic for another book. My search for Gondwana is over.

I stand next to the Scott monument and look out over the Otago Harbour, sweeping out towards the heads like a broad blue paint stroke. The peninsula opposite is a patchwork of pale green farmland. Once, those hillsides were clothed in an ancient forest of podocarps and southern beech trees. Now, the hills of the Otago Peninsula are largely bare of trees, save for rows of introduced ones, pine and *Macrocarpa*.

But in every nook and cranny of this land, in every shaded gully where stock can't graze, and on every steep hillside where crops can't be grown, plants riot towards the sun. We call them weeds: gorse, broom, briar, ragwort, thistles ... the list in New Zealand is endless. Below me, the embankment that plunges down to the railway line is a chaos of weeds. Introduced trees, sycamores and *Eucalyptus*, crowd the slope, all of them festooned with invasive creepers like banana passionfruit and old man's beard. Gorse smothers the country, growing in great fists alongside the railway line and up along the opposite hill.

I look out on all of this and realize we haven't really tamed this land. We may have upset the old ways and changed the flora beyond recognition, but the voracity for growth that plants have is still inherent in the soil. Despite our very best efforts at destroying it, containing it and controlling it, the wildness is still there, ceaselessly clamouring for its inevitable chance.

*Glossopteris* was just one of the millions of forms of life that have lived on this planet and then vanished. It survived for 50 million years — that's 1000 times as long as human beings have existed. But then, like 99.9 per cent of everything that has ever lived on Earth, it was gone. The ground literally shifted, and suddenly it could no longer survive here.

I gaze at Scott's memorial, contemplating this. What mean these stones? Captain Scott sailed to the Antarctic looking to answer that question. He found only disappointment and death. Nothing on this planet is sacred. In an evolutionary sense, we are all disposable.

And as I look out over this 'weed' infested land, I realize that if we humans were to vanish from this Earth, plants would inherit it, completely and immediately.

This, of course, should come as no surprise. It was theirs to begin with.

# EPILOGUE: 250 MILLION YEARS IN THE FUTURE

We float in space, high above a new world. Below us, a planet so familiar, yet utterly alien, turns in the void. From this angle it's a sea world, entirely composed of water — an ocean that stretches from horizon to horizon.

As we descend we can observe this ocean's textures and moods. We recognize her long, lazy swell, the calm rhythm of which belies her stormy spirit. We understand her waters, an intense, deep blue in the tropics, steely grey in the polar regions. A huge storm is building at the equator, and we watch it slowly spiralling across the face of the world. With an almost endless warm ocean to draw upon, and no land to break its progress, it grows with unconstrained power.

Now, as the planet turns, Earth's other side is revealed and we see where all the land has gone. A huge conglomeration of continents is once again formed, stretching from ice-free pole to ice-free pole.

The Atlantic has vanished, closed up by the collision between South America and Africa. The Mediterranean too has gone, and what was Africa now piles into Europe, a great mountain range seaming the continent, running from the remains of Europe, across into Asia and emerging at the Panthalassic Coast. This is Pangaea Proxima, the next supercontinent. From pole to pole, forests sweep across it. There is no sign of humanity here; where once our cities festooned the fringes of the continents, there is now only wildness.

We descend through the clouds and land on its shore at the fringes of a towering forest. The sea churns at the base of a low cliff. Far out to sea, squadrons of strange flying animals that might or might not be birds, tens of thousands of them, thread through the sky in enormous, continuous V formations that stretch as far as we can see.

We turn to face the forest. Hulking trees stand sentry at the entrance, a thick wall of leaves forming the canopy. Creeping plants are draped across trees and shrubs. As we move towards them, scarlet carnivorous flowers swivel on their stems, turning slowly towards us. A low whisper passes through the leaves above.

A sure and thrilling knowledge enters our consciousness. This forest knows we're here.

We take our first tentative steps into an unknown world.

# GLOSSARY

**Ammonite** Swimming molluscs with coiled shells, the biggest of which may have grown to more than 3 metres (10 feet) in diameter. Ammonites went extinct at the end of the Cretaceous period.

**Amphibian** A class of animals that need water, or moist environments, to survive.

**Angiosperm** A term that describes all the flowering plants.

*Anteosaurus* A carnivorous dinocephalian therapsid of the Mid Permian, known from fossils in South Africa.

**Anthracite** The highest grade of coal, having a carbon content greater than 86 per cent.

*Araucaria* A genus of conifer trees that includes the monkey puzzle, kauri and Norfolk Island pine.

*Araucarioxylon* A genus of fossil conifers, described solely from the structure of the wood.

*Archaeopteris* An extinct, spore-producing tree that probably formed Earth's first forests during the Devonian period.

**Asthenosphere** The portion of the Earth's upper mantle which is ductile, allowing tectonic plates to move horizontally above it.

**Auxin** Plant hormones that control growth. By elongating cells in stems, auxin can alter the direction of growth.

**Avalonia** A Paleozoic continent. During the Silurian, Avalonia collided with Baltica and Laurentia to form Laurussia. The remnants of Avalonia are today found in western Europe, Great Britain, Canada and the United States.

———————

**Baltica** A continent which collided with Avalonia and Laurentia to form Laurussia during the Paleozoic era.

**Basalt** A dark, fine-grained rock formed when lava cools rapidly at the planet's surface. Basalt is the most common volcanic rock on Earth and forms the basis of oceanic crust.

**Basin** (Geology) A large, low-lying region into which sediments drain.

**Beacon Supergroup** A geological unit of sedimentary rock in Antarctica comprising sandstones, shales and conglomerates. Paleozoic-Mesozoic age.

**Beardmore Glacier** A 200-kilometre-long (125-mile) glacier on Antarctica that descends from the Antarctic Plateau to the Ross Ice Shelf. One of the largest valley glaciers on Earth.

**Beaufort Group** A stratigraphic section within South Africa's Karoo Supergroup which is rich in vertebrate fossils and records the Permian–Triassic boundary.

**Bifurcating** (Plants) The division of a plant stem into two branches.

**Biome** A large area characterized by the types of climate, ecosystems and flora and fauna it contains. There are seven major biomes in the modern world — tropical and temperate rainforest, desert, tundra, taiga, grassland and savanna.

**Bituminous coal** The second-highest grade of coal (after anthracite), having a carbon content of 76 to 86 per cent.

**Bivalve** A class of marine and freshwater molluscs characterized by having two hinged shells. Modern examples include mussels and clams.

**Brachiopod** A phylum of marine animals that have hinged shells and which therefore superficially resemble bivalves. Brachiopods were heavily impacted by the Permian extinction, after which bivalves rose to prominence.

**Bryozoan** A phylum of invertebrate aquatic animals that form intricate colonies from calcium carbonate.

---

**Cambium** The thin layer of living tissue in the stems and roots of plants which produces secondary xylem and phloem.

**Cambrian explosion** A period of time during the Cambrian when almost all the major modern phyla of animal suddenly appear in the fossil record. Prior to the Cambrian, only unicellular and relatively simple multicellular animals existed on Earth. The Cambrian explosion marked a spectacularly rapid diversification of complex animals. Some researchers have argued that these events were in part triggered by the formation of the Trans-Gondwanan mountains.

**Captorhinids** A family of small lizard-like reptiles that existed in the Permian period.

**Carbon cycle, the** The regular cycling, through chemical reactions and tectonic activity, of carbon atoms through rocks, soil, the ocean and living organisms.

**Chalk** A soft, white sedimentary rock formed under the sea by the settlement of microscopic plankton.

**Chloroplast** Organelles that exist inside plant cells and which conduct photosynthesis.

**Cimmerian terranes** A string of landmasses that broke away from northern Gondwana and combined with Laurasia during the Permian and Triassic periods. Remnants of these terranes today form parts of Turkey, Iran, Afghanistan, Pakistan, Tibet, China, Myanmar, Thailand and Malaysia.

**Columbia** A supercontinent thought to have existed 2.5 to 1.5 billion years ago.

**Comparative anatomy** The study of similarities and differences in the body structures of living and extinct animals.

**Conifer** The group of cone-bearing plants (gymnosperms) that includes pines, podocarps and araucaria trees.

**Continental crust** The parts of the Earth's crust that form the continents. Formed largely from granite, continental crust is less dense than oceanic crust and typically much thicker. It rides atop oceanic crust in the way an iceberg floats on water.

**Continental drift** The hypothesis that through time continents have moved, relative to each other, across the Earth's surface.

**Convection currents** (Plate tectonics) Currents that transfer heat generated by radioactive decay from deep in the Earth's interior to the asthenosphere.

**Coral, rugose** An order of solitary corals that existed prior to the Permian extinction.

**Coral, tabulate** An order of colonial corals that existed prior to the Permian extinction.

*Cordaites* An order of gymnosperms that were common in the Carboniferous and Permian periods.

**Core, inner** The innermost part of planet Earth — a solid metallic ball made mostly of iron.

**Core, outer** A fluid layer in the Earth's structure that sits above the inner core. The outer core is composed mostly of iron and nickel.

**Craton** An old portion of continental crust that has survived the splitting apart and merging of continents for hundreds of millions of years.

**Crinoid** Filter-feeding marine animals.

**Crust** The outermost layer of the Earth, which is composed of oceanic and continental crust.

**Cyanobacteria** Photosynthesizing microorganisms that live in water. Also known as blue-green algae.

**Cycad** Seed-producing gymnosperms that superficially resemble palms but are more closely related to conifers. An ancient group with its origins in the Permian period.

**Cynodont** A group of therapsids that thrived in the late Permian. Cynodonts survived the Permian mass extinction and diversified into many lineages, including mammals. Humans are cynodonts.

*Daptocephalus* **Assemblage Zone** The uppermost Permian unit of the Beaufort Group — a layer of rock named for the prevalence of fossils of the dicynodont *Daptocephalus*.

**Diapir** A geologic intrusion in which mobile material (e.g. magma) is forced into brittle overlying rock.

*Dicroidium* A genus of seed fern that flourished on Gondwana after the Permian extinction.

**Dicynodont** A group of therapsids that thrived during the Permian period. Aside from a pair of tusks, dicynodonts lacked teeth and ground their food with a horny beak.

**Dinocephalians** A group of therapsids that flourished during the Mid Permian period.

**Dinosaur** A group of sauropsids that thrived during the Mesozoic Era.

**Dispersal** (Biogeography) The movement of species from one place to another.

**Echinoderms** A group of invertebrates characterized by their hard, spiny coverings. Living examples include sea urchins and starfish.

**Eclogite** A metamorphic rock found deep in the Earth's crust, formed by the intense pressure and heat that mafic rock experiences when it is subducted to great depth.

**Ecoregion** An area in which ecosystems are generally similar.

**Embryo** (Plant) The part of a seed or bud that contains the earliest form of the plant's roots, stems and leaves.

**Enlightenment, the age of** An intellectual movement in 17th and 18th century Europe, during which many old ideas, customs and traditions were cast aside in favour of scientific enquiry and individual reasoning.

*Eodicynodon* **Assemblage Zone** The oldest rock unit of the Beaufort Group in South Africa's Karoo, named for the prevalence of fossils of the dicynodont *Eodicynodon*.

**Epicormic shoots** Buds on a plant's stem that produce shoots after stress, such as a fire.

**Epiphyte** A plant that attaches itself to the trunk or stems of another plant.

**Eukaryotic cell** A cell which contains a nucleus and other organelles (as opposed to prokaryotic cells, which are more ancient, and do not).

**Exoskeleton** The rigid eternal structure that supports the internal organs of some invertebrates, such as insects.

_____

**Fault** A fracture in rock. The largest faults form the boundaries between tectonic plates.

**Fern** An ancient group of spore-producing plants that have existed since at least the Devonian period.

**Form genus** A taxonomic classification used for fossils that exhibit similarities in structure, but which cannot reliably be assigned to any particular species or genera, such as parts of a plant, like roots and wood.

**Fructification** The reproductive organs of a plant.

_____

**Gametes** An organism's reproductive cells.

**Germination** The sprouting of a seed or spore.

*Gigantopteris* A genus of plant that thrived in the Permian before vanishing by the early Triassic.

**Ginkgo** An order of trees that thrived in the Permian period, and which is today represented by a single species, *Ginkgo biloba*.

*Glossopteris* A genus of seed ferns that thrived on Gondwana before becoming extinct at the end of the Permian period.

**Gondwana (aka Gondwanaland)** A supercontinent that existed from the Neoproterozoic era until the Jurassic period.

**Gorgonopsids** Sabre-toothed, carnivorous therapsids that were dominant predators during the late Permian period.

**Granite** Igneous rock formed by the extrusion of slowly cooling magma into the Earth's crust. Granite forms the basis of continental crust.

**Graptolite** Colony-forming marine animals of the Paleozoic era.

**Greywacke** Sedimentary rock formed from mud and sand at the edge of continental shelves.

**Gymnosperm** A term for plants that do not have flowers and produce 'naked seeds'. Conifers, cycads and ginkgos are all gymnosperms.

**Gypsum** A soft, sulphate mineral.

_____

**Horsetail** An ancient family of plants that dominated the understorey of Permian forests, today represented by just one genus, Equisetum.

**Hot spot** An area of the Earth's crust that sits above a mantle plume. Volcanic islands like Hawaii and the

Cape Verde Islands are thought to be produced at mantle hot spots.

**Hyphae** Branching filaments that make up the mycelium of a fungus.

---

**Igneous rock** Rock formed by the cooling of magma or lava.

**Ionization** The process by which electrically neutral atoms are converted to electrically charged atoms.

---

**Karoo Supergroup** A stratigraphic unit in South Africa spanning a period of 120 million years, between the Carboniferous and Jurassic periods.

**Kenorland** A supercontinent thought to have existed some 2.5 billion years ago.

---

**Large Igneous Province (LIP)** A huge accumulation of igneous rocks on the Earth's surface, formed by prolonged and massive outpourings of lava. LIPs are thought to form over mantle plume hot spots.

**Laurasia** A supercontinent comprising the northern part of Pangaea. Laurasia broke apart during the Mesozoic to form Europe, Asia, North America and other Northern Hemisphere landmasses.

**Laurentia** A continental craton that today forms most of North America. Laurentia collided with Avalonia and Baltica during the Silurian period to form Laurussia.

**Laurussia** A supercontinent formed from the collision of Avalonia, Baltica and Laurentia during the Paleozoic era. Laurussia collided with Gondwana during the Carboniferous period to form Pangaea.

**Lava** Magma that erupts onto the Earth's surface.

*Lepidodendron* Giant clubmoss trees that dominated forests during the Carboniferous period.

**Lignite** The lowest grade of coal, having a carbon content of 25 to 35 per cent.

**Limestone** A marine sedimentary rock formed from calcite and aragonite. Often rich in fossils.

**Lithosphere** The outermost shell of the Earth, consisting of the crust and the uppermost solid mantle. The lithosphere sits atop the ductile asthenosphere.

**Liverwort** A division of non-vascular plants whose ancestors were among the earliest land plants.

**Lycopod** The group of plants known as clubmosses. While diminutive today, in the Carboniferous they formed giant trees.

**Lystrosaurus** A dicynodont that thrived across the world in the aftermath of the Permian extinction.

_____

**Magma** Molten rock.

**Magma chamber** The area beneath a volcano in which magma is stored prior to an eruption.

**Mammoth** An elephant-like animal that inhabited North America between the Miocene and Pleistocene epochs.

**Mantle** The portion of the Earth's interior that is above the core and beneath the crust.

**Mantle plume** A theoretical phenomenon in which superheated material rises through the Earth's mantle to the base of the crust.

**Mastodon** An elephant-like animal that inhabited North America between the Miocene and Pleistocene epochs.

**Metamorphic rock** Rock that was once sedimentary or igneous rock but which has been chemically altered by heat and pressure to form a new type of rock.

**Mica** A mineral whose crystals split easily to form flat sheets that can cause rocks to appear 'sparkly'.

**Mid-ocean ridge** A chain of undersea mountains formed by a rifting of oceanic crust, which allows magma to erupt on the sea floor. Mid-ocean ridges are the source of new oceanic crust.

**Midrib** The central rib on a leaf.

**Mirovia** The ocean that surrounded the supercontinent Rodinia.

**Moa** An extinct genus of flightless ratites from New Zealand, the biggest of which were among the largest birds ever to exist.

**Moho Discontinuity** The boundary between the Earth's crust and the mantle.

**Mollusc** A phylum of invertebrates comprising 85,000 extant species and perhaps 100,000 more fossil species.

**Monsoon** A weather system characterized by seasonally reversing winds and corresponding changes in precipitation.

**Mosasaur** A large carnivorous marine reptile that existed in the Cretaceous period.

**Moschops** A herbivorous Mid Permian dinocephalian therapsid.

**Mycelium** The vegetative part of a fungus, which exists underground. The mycelium consists of a tangled network of hyphae.

**Myrtaceae** A family of plants common in the Southern Hemisphere. Living examples include Eucalyptus, pōhutakawa, allspice and guava.

*Nothofagus* A genus of tree and shrubs found only in the Southern Hemisphere.

**Ocean trench** A deep depression in the sea floor, formed at a subduction zone, where oceanic crust is being pulled down into the Earth's mantle.

**Oceanic crust** The parts of the Earth's crust that underlie the sea. Oceanic crust is made of basaltic rocks and is denser and generally thinner than continental crust. It is formed at mid-ocean ridges and destroyed at subduction zones.

**Organelle** Structures inside cells which perform specialized tasks, e.g. chloroplasts.

**Paleo-Tethys Ocean** A sea which lay between Laurasia and Gondwana during the Paleozoic era.

**Pangaea** A supercontinent comprised of Laurasia in the north and Gondwana in the south, which existed between the Early Permian period and the Jurassic period.

**Panthalassic Ocean** The ocean that surrounded Pangaea.

*Pareiasaurus* A large, herbivorous sauropsid that existed in the Permian.

**Patagonia** The southern portion of South America, incorporating parts of Argentina and Chile.

**Peat** An accumulation of partially decayed vegetation, which forms bogs. The first stage in coal formation.

**Pegmatite** An igneous rock that has cooled slowly and is thus formed of very large crystals.

**Peridotite** An igneous rock found deep in the earth's interior, the dominant rock type of the Earth's upper mantle.

**Permineralized** A process of fossilization in which minerals fill the spaces left by the internal casts of organisms.

**Phloem** Plant stem tissue through which food is distributed from the leaves to the rest of the plant. Secondary phloem forms bark.

**Photon** An elementary particle, the fundamental particle of light.

**Photosynthesis** The chemical reaction by which plants and some bacteria manufacture their food from sunlight and water.

*Phyllotheca* A species of horsetail common in the Permian period, a plant that is thought to have formed pastures on which herbivores grazed.

**Plesiosaur** A carnivorous marine reptile that lived during the Jurassic period.

*Pleuromia* A species of lycophyte that dominated vegetation during the early Triassic.

**Podocarp** A genus of conifers found mostly in the Southern Hemisphere.

**Pollen** A powdery substance that contains the male gametes of plants.

*Productus* A genus of brachiopod common in the Carboniferous and Permian periods.

*Proteaceae* A family of flowering plants found mostly in the Southern Hemisphere.

**Proton** A positively charged particle found inside an atomic nucleus.

*Prototaxites* A genus of fungi that lived between the Ordovician and Devonian periods. Some examples are thought to have grown up to 8 metres (26 feet) in height.

---

**Radioactive decay** The process by which an unstable atomic nucleus loses energy, by emitting particles.

**Rheic Ocean** An ocean which separated Laurussia and Gondwana before its closure drew the two continents together to form Pangaea.

**Rift valley** A lowland created by a geological rift, where two tectonic plates are pulling apart.

**Rock cycle** The geological cycle that describes the transition between igneous, sedimentary, and metamorphic rocks. For example, igneous rocks produced by subduction zone volcanoes are eroded to form sedimentary rock on the seabed, which is then subducted and altered by heat and pressure to form a metamorphic rock. This is then chemically recycled into the mantle and may once again be erupted at the surface.

**Rodinia** A supercontinent that existed between 1 billion and 650 million years ago.

*Reduviasporonites* A microfossil, though to be of a fungi, which occurs in great quantities in the post-Permian fossil record.

**Ross Ice Shelf** A region of sea ice in the Ross Sea forming an unbroken expanse, the largest ice shelf in Antarctica, having an area of over half a million square kilometres (193,000 square miles).

---

**Sandstone** A type of sedimentary rock formed from the deposition of sand-sized grains.

**Scientific revolution, the** A series of events in the 16th and 17th century that marked the emergence of modern science.

**Seafloor spreading** The rifting apart of tectonic plates.

**Sedimentary rock** A class of rocks formed by the accumulation of material or organic particles.

**Seed** A plant embryo enclosed in a protective covering.

**Shale** A type of fine-grained sedimentary rock formed on the seabed or in floodplains, lagoons and deltas.

**Slab** The portion of a tectonic plate that is subducted.

**Slab graveyard** An area at the base of the mantle at which subducting slabs collect before being chemically recycled into the mantle. Slab graveyards may be a source of combustion for mantle plumes.

**Snowball Earth** A theory that at times in the Earth's history, the planet has been entirely frozen.

**Sphenophyte** An ancient group of vascular plants, today represented by just one genus of horsetail, *Equisetum*.

**Spore** A unit of sexual reproduction in some plants.

**Stomata** Tiny openings in plant leaves and stems that facilitate gas exchange. Carbon dioxide enters the plant through the stomata. Stomata open and close to control water loss.

**Strata** Layers of sedimentary rock.

**Subduction** The process by which oceanic lithosphere is recycled into the mantle.

**Subduction zone** An area in which oceanic lithosphere is being subducted beneath continental crust. The subduction process destroys oceanic crust and leads to the formation of volcanic arcs.

**Supernova** The super-powerful explosion of a star.

**Superposition, Law of** A principle of stratigraphy stating that younger rocks overlie older rocks in a stratigraphic sequence.

**Synapsid** One of the two major groups of animals that evolved from early amniotes (the other being sauropsids). Synapsids include the therapsids, a dominant group of land animals during the Permian. After the Permian extinction, synapsids were devasted by the Permian extinction, although some lineages survived. After the Cretaceous/Paleogene event that terminated the dinosaurs' reign, synapsids again rose to prominence. Among them were our own ancestors.

---

**Taproot** A large, central plant root from which other roots sprout.

**Terrane** A fault-bounded region with a distinctive geological structure and history.

**Tethys Ocean** An ocean that lay between Laurasia and Gondwana from the late Paleozoic to the Jurassic, formed as the Cimmerian terranes drew northward to close the Paleo-Tethys Ocean.

**Therapsid** The group of synapsids that includes, among other lineages, the dinocephalians, gorgonopsids, and cynodonts. Humans are therapsids.

**Thermodynamics** A branch of physics that investigates heat, temperature and energy.

**Tillite** Unsorted sediment carried and deposited by a glacier.

**Transform fault** A type of fault in which two tectonic plates slide past one another.

**Trilobite** A group of marine arthropods that were one of the most successful groups of early animals. Trilobites appeared in the Cambrian and became extinct at the end of the Permian period.

———————————

**Ultraviolet light** A form of electro-magnetic radiation with a shorter wavelength than visible light. Excessive exposure to UV light can cause cancer and skin damage.

**Uniformitarianism** The theory that geological processes operating today are continuous and have occurred throughout Earth's history.

**Ur** A supercontinent thought to have formed 3.1 billion years ago.

**Vascular system** (Plants) A system of tissues, strengthened by lignin,

for transporting water and minerals around the plant.

**Venation** The veins on a leaf.

**Vertebra** One of the bones that forms the spine of a vertebrate animal.

*Verterbraria* A form genus to describe the root system of certain fossil plants. The root systems of *Glossopteris* fall in this genus.

**Vicariance** The separation of a population of plants or animals by a geographical barrier, such as a mountain range or an ocean, which results in the population splitting into two or more species.

**Volcanic Arc** A chain of volcanoes that forms above a subduction zone (e.g. the Andes).

*Voltziopsis* A conifer that thrived in the early Triassic.

———————————

*Winteraceae* A family of tropical trees and shrubs mostly found in the Southern Hemisphere.

———————————

**Xylem** Vascular tissue in plants that conducts water and dissolved nutrients upwards from the roots. Secondary xylem forms wood.

# ACKNOWLEDGMENTS

I am greatly indebted to Ewan Fordyce, whose inspirational teaching at the University of Otago first set me on the road to Gondwana.

I would like to thank my wife Emma and Jim and Mary Ann Morris for their love and support as I wrote this book.

I would also like to thank Bari Cariglino, Rose Prevec, Stephen McLoughlin, Greg Retallack, Peta Hayes, Paul Kenrick, Aly Baumgartner, Nick Mortimer, Sam van der Weerden, Cameron Brideoake, Emma Edwards, Hamish Campbell, Daphne Lee, Sophie White, Lemuel Lyes, Marcus Richards, Lloyd Davis and Chuck Landis.

Thank you to Paulina Barry for her stunning original illustrations.

I'd especially like to acknowledge Geoff and Trish Dalbeth for their financial support. The travel in this book was made possible by a grant from the Shackleton Scholarship Fund, Falkland Islands. I am very grateful to the Fund for this opportunity.

Finally, thanks to everyone at Exisle Publishing for believing in this project.

# ENDNOTES

## Prologue

1. Fiennes, R., 2003, *Captain Scott*, Hodder and Stoughton, London, pp. 192–3, 339–77.

2. Scott, R.F., 2005, *Journals*, Edited by Stone, M., Oxford University Press, Oxford, p. 376, p. 392, pp. 394–412.

3. Harrington, H.J. and Speden, I., 1962, 'Section through the Beacon Sandstone at Beacon Height West, Antarctica', *New Zealand Journal of Geology and Geophysics* 5 (5), pp. 707–17.

4. Anthony, J., 2012, *Hoosh: Roast penguin, scurvy day, and other stories of Antarctic cuisine*, University of Nebraska Press, Lincoln, p. 55.

5. McDougall, I. and Coombs, D.S., 1973, 'Potassium-argon ages for the Dunedin volcano and outlying volcanics', *New Zealand Journal of Geology and Geophysics*, 16 (2), pp. 179–88.

## Chapter 1

1. Barr, J., 1984–85, 'Why the world was created in 4004 BC: Archbishop Ussher and biblical chronology', *Bulletin of the John Rylands University Library of Manchester*, 67, pp. 603–607.

2. Rudwick, M., 2016, *Earth's Deep History: How it was discovered and why it matters*, University of Chicago Press, Chicago, pp. 15–16.

3. Oldstone-Moore, C., 2021, *Commoner's' Life*, Wright State University, www.wright.edu/~christopher.oldstone-moore/richessay.htm

4. Wolchover, N., 2012, 'Why it took so long to invent the wheel', https://www.livescience.com/18808-invention-wheel.html

5. Desmond, A.J., 1975, 'The discovery of marine transgression and the explanation of fossils in antiquity', *American Journal of Science*, 275 (6), pp. 692–707.

6. Baucon, A., 2010, 'Leonardo da Vinci, the founding father of Ichnology', *Palaios* 25 (5/6), pp. 361–67.

7. Cutler, A.H., 2009, 'Nicolaus Steno and the problem of deep time. The revolution in geology from the Renaissance to the Enlightenment 203', *Geological Society of America*, pp. 143–48.

8. Rudwick, M., 2016, pp. 40–41.

9. Rudwick, M., 2016, pp. 66–8.

10. Egerton, F.N., 2007, 'A history of the ecological sciences, part 24: Buffon and environmental influences on animals', *Bulletin of the Ecological Society of America*, 88 (2), pp. 146–59.

11. Rudwick, M., 2005, *Bursting the Limits of Time: The reconstruction of geohistory in an age of revolution*, University of Chicago Press, Chicago, pp. 443–534.

12. Dean, D., 1992, *James Hutton and the History of Geology*, Cornell University Press, New York, p. 6.

13. Rudwick, M., 2016, pp. 103–109.

14. Preaud, T., et al., 1997, *The Sèvres Porcelain Manufactory: Alexandre Brongniart and the triumph of art and industry, 1800–1847,* published for the Bard Graduate Center for Studies in the Decorative Arts, Yale University Press, New Haven, pp. 1–416.

15. Rudwick, M., 1996, 'Cuvier and Brongniart, William Smith, and the reconstruction of geohistory', *Earth Sciences History: Journal of the History of the Earth Sciences Society* 15 (1), pp. 25–36.

16. Li, Z.X., et al. 2008, 'Assembly, configuration, and break-up history of Rodinia: A synthesis', *Precambrian Research*, 160 (1), pp. 179–210.

## Chapter 2

1. Arposio, A., 2018, *Collected Writings on the Awabakal People*, Miromaa Aboriginal Language and Technology Centre, Newcastle, p. 28.

2. Donnelly, M., 2017, 'Convict escapees lit up Newcastle coal finds', *Daily Telegraph*, 18 September.

3. Orem, W.H., and Finkelman, R.B., 2003, *Coal Formation and Geochemistry*, Treatise on Geochemistry, p. 407.

4. Fernihough, A., 2021, 'Coal and the European industrial revolution', *Economic Journal*, 131 (635), pp. 1135–49.

5. Doctor, V., 2012, 'Coal dust in the Tagore Album', *Economic Times*, 20 September.

6. Robins, N., 2002, 'Loot: In search of the East India Company, the world's first transnational corporation', *Environment and Urbanization*, 14 (1), pp. 79–88.

7. Asiatic Society of Bengal, 1842, volume 11, part 2, *Journal of the Asiatic Society of Bengal*, pp. 736–7.

8. Greb, S.F., 2013, 'Coal more than a resource: Critical data for understanding a variety of Earth-science concepts', *International Journal of Coal Geology*, 118 (October), pp. 15–32.

9. Stafleu, F., 1966, 'Brongniart's Histoire Des Végétaux Fossiles', *Taxon* 15 (8), pp. 320–24.

10. Rudwick, M., 2016, p. 137.

11. Brongniart, A., 1838, 'Reflections on the nature of the vegetables which have covered the surface of the earth at different periods of its formation', *Magazine of Natural History, and Journal of Zoology, Botany, Mineralogy, Geology and Meteorology*, pp. 2–13.

12. Andrews, H.N., 1980, *The Fossil Hunters: In search of ancient plants*, Cornell University Press, Ithaca, p. 412.

13. Brongniart, A., 1828. *Histoire des Végétaux Fossiles ou Recherches Botaniques et Géologiques sur les Végétaux Renfermés dans les Diverses Couches du Globe*, Chez G. Dufour et Ed. d'Ocagne, Paris, pp. 223–4.

14. Stone, P., 2010, 'The Geology of the Falkland Islands', *Deposits Magazine*, 23, pp. 38–43.

15. Darwin, C., 1846, *Geological Observations on South America: Being the third part of the geology of the voyage of the Beagle, under the command of Capt. Fitzroy, R.N. during the years 1832 to 1836*, Smith, Elder and Company, London, pp. 51–69.

16. Rushton, A., Stone, P., 2011, 'Two notable fossils finds in East Falkland: A starfish and a large trilobite', *Falkland Islands Journal*, 9 (5), pp. 5–13.

17. Cohen, A., 2000, 'Mr Bain And Dr Atherstone: South Africa's pioneer fossil hunters', *Journal of the History of the Earth Sciences Society*, 19 (2), pp. 175–91.

18. Owen, R., 1845, 'Report on the reptilian fossils of South Africa: Part I — Description of certain fossil crania, discovered by A.G. Bain, Esq., in sandstone rocks at the south-eastern extremity of Africa, referable to different species of an extinct genus of Reptilia (*Dicynodon*), and indicative of a new tribe or sub-order of Sauria', *Transactions of the Geological Society of London*, S2–7 (1), pp. 59–84.

19. Davenport, J., 2016, 'The father of South African geology', *Mining Weekly*, 21 October, www.miningweekly.com/article/the-father-of-south-african-geology-2016-10-21.

20. Johnson, M.R., Van Vuuren, C.J., Hegenberger, W.F., Key, R., and Show, U., 1996, 'Stratigraphy of the Karoo Supergroup in Southern Africa: An overview', *Journal of African Earth Sciences*, 23 (1), pp. 3–15.

21. Chakrabarti, P., 2019., 'Gondwana and the Politics of Deep Past', *Past & Present*, 242 (1), pp.119–53.

22. Blunt, J.T., 1803, *Asiatick Researches Volume 7*, Bengal Military Orphans Press, Kolkata, p. 96.

23. Sleeman, W.H., 1844, *Rambles and Recollections of an Indian Official*, J. Hatchard and Son, London, volume 1, p. 6.

24. Hislop, S., and Hunter, R., 1854, 'On the geology of the neighbourhood of Nágpur, central India', *Quarterly Journal of the Geological Society of London*, 10 (1–2), pp. 470–73.

25. Smith, A.J., 1963, 'Evidence for a Talchir (lower Gondwana) glaciation; Striated pavement and boulder bed at Irai, central India', *Journal of Sedimentary Research*, 33 (3), pp. 739–50.

26. Saxena, A., Singh, J.K., and Goswami, S., 2014, 'Advent and decline of the genus *Glossopteris brongniart* in the Talcher coalfield, Mahanadi basin, Odisha, India', *The Paleobotanist*, 63 (January), pp. 157–68.

27. Chakrabarti, P., 2020, *Inscriptions of Nature: Geology and the naturalization of antiquity*, JHU Press, Baltimore, pp. 161–92.

## Chapter 3

1. Armstrong, T., 1971, 'Bellingshausen and the discovery of Antarctica', *The Polar Record*, 15 (99), pp. 887–9.

2. Paine, Lincoln P., 2000, *Ships of Discovery and Exploration*, Houghton Mifflin, Boston, pp. 139–40.

3. Seward, A.C., 1912, 'Sir Joseph Hooker and Charles Darwin: The history of a forty years' friendship', *The New Phytologist*, 11 (5/6), pp. 195–206.

4. Endersby, J. 2001. 'Joseph Hooker: The making of a botanist', *Endeavour*, 25 (1), pp. 3–7.

5. Hooker, J.D., 1853, *Introductory Essay to the Flora of New Zealand*, Lovell Reeve, London, p. xxi.

6. Darwin, C., personal letter, 5 November 1845, www.darwinproject.ac.uk/

7. Darwin, C., personal letter, 13 June 1850, www.darwinproject.ac.uk/

8. Darwin, C., personal letter, 24 April 1855, www.darwinproject.ac.uk/

9. Williams, N., 2008, 'Darwin celebrations begin', *Current Biology: CB*, 18 (14), pp. 579–80.

10. Şengör, A.M.C., 2014, 'Eduard Suess and global tectonics: An illustrated "short guide"', *Austrian Journal of Earth Sciences*, 107/1, pp. 6–82.

11. Şengör, A.M.C., 2015, 'The Founder of Modern Geology Died 100 Years Ago: The Scientific Work and Legacy of Eduard Suess', *Geoscience Canada* 42 (2), p. 181.

12. Suess, E., 1904, *The Face of the Earth (Das Antlitz de Erde)*, Clarendon Press, Oxford, pp. 440, 417, 625, 633.

13. Darwin, C., 1859, *On the Origin of Species by Means of Natural Selection, Or, The Preservation of Favoured Races in the Struggle for Life*, reprinted 2009, Vintage, London, pp. 758–9.

14. Burchfield, J.D., 1990, *Lord Kelvin and the Age of the Earth*, University of Chicago Press, Chicago, p. 32.

15. Radvanyi, P., and Villain, J., 2017, 'The discovery of radioactivity', *Comptes Rendus Physique*, 18 (9), pp. 544–50.

16. Boltwood, B.B., 1907, 'Ultimate disintegration products of the radioactive elements: Part II, disintegration products of Uranium', *American Journal of Science*, 23 (February), p. 78.

17. Senter, P., 2013, 'The age of the Earth and its importance to biology', *American Biology Teacher*, 75 (4), pp. 251–6.

18. Tonnessen, J., and Johnsen, A., 1982, *The History of Modern Whaling*, C. Hurst and Company, London, pp 147–50.

19. Feldmann, R.M., and Woodburne, M., 1988, 'Geology and paleontology of Seymour Island Antarctic Peninsula', *Geological Society of America*, 169, pp. 1–6.

20. Nordenskjold, O., 1904, 'The Swedish Antarctic expedition, 1902–03', *Bulletin of the American Geographical Society*, 36 (1), pp. 22–9.

21. Aldiss, D.T., and Edwards E.J., 1999, 'The geology of the Falkland Islands', *British Geological Survey*, p. 50.

22. Fiennes, R., 2003, *Captain Scott*, Hodder and Stoughton, London, pp. 192–3.

23. Chaloner, W., and Kenrick, P., 2015, 'Did Captain Scott's Terra Nova Expedition discover fossil Nothofagus in Antarctica?' *The Linnean*, 31 (2), pp. 11–17.

24. Crane D, 2006, *Scott of the Antarctic*, Harper Perennial, New York, p. 386.

25. Fiennes, R., 2003.

26. Greene, M.T., 2015, *Alfred Wegener: Science, exploration, and the theory of continental drift*, JHU Press, Baltimore, pp. 286-313.

27. Coniff, R., 2012, 'When Continental Drift was Considered PseudoScience.' *Smithsonian Magazine*, www.smithsonianmag.com/science-nature/when-continental-drift-was-considered-pseudoscience-90353214/

28. Wegener, A., 1966, *The Origin of Continents and Oceans*, Dover Publications, New York, pp. 98–142.

29. Le Grand, H.E., 1986, 'Steady as a Rock: Methodology and moving continents', in *The Politics and Rhetoric of Scientific Method: Historical studies*, Springer, Dordrecht, pp. 97–138.

30. England, R., Kearey, P., Klepeis K.A., and Vine, F., 2009, 'Global tectonics', *Marine Geophysical Research*, 30 (4), pp. 293–4.

31. Wilson, R.W., et al., 2019, 'Fifty years of the Wilson cycle concept in plate tectonics: An overview', *Geological Society, London, Special Publications*, 470 (1), pp. 1–17.

32. Squire, R., et al., 2006, 'Did the Transgondwanan supermountain trigger the explosive radiation of animals on Earth?' *Earth and Planetary Sciences Letters*, 250 (1–2), pp. 116–33.

## Chapter 4

1. Le Grand, H.E., 1986.

2. Du Toit, A.L., 1937, *Our Wandering Continents*, Oliver and Boyd, Edinburgh, pp. 1–36.

3. Frankel, H., 1978, 'Arthur Holmes and continental drift', *British Journal for the History of Science*, 11 (2), pp. 130–50.

4. Longwell, C., 1944, 'Some thoughts on the evidence for continental drift', *American Journal of Science*, 242, pp. 218–31.

5. Frankel, H., 2012, *The Continental Drift Controversy*, Cambridge University Press, Cambridge, pp. 264–38.

6. Sahni, B., 1936, *Wegener's Theory of Continental Drift in the Light of Palaeobotanical Evidence*, Associated Printers, Chandigarh, pp. 319–22.

7. Şengör, A.M.C., 2014.

8. Şengör, A.M.C., 2015, pp. 181–246.

9. Dudley, W., 1986, 'Gold from the sea', *Journal of Geological Education*, 34 (1), pp. 4–6.

10. Picard, M.D., 1989, 'Harry Hammond Hess and the theory of sea-floor spreading', *Journal of Geological Education*, 37 (5), pp. 346–9.

11. Vine, F.J., and Matthews, D.H., 1963, 'Magnetic anomalies over oceanic ridges', *Nature*, 199 (4897), pp. 947–9.

12. Arkani-Hamed, J., 2017, 'Formation of a solid inner core during the accretion of Earth', *Journal of Geophysical Research, [Solid Earth]*, 122 (5), pp. 3248–85.

13. Keary, P., Kelpeis, K., and Vine, F., 2009, *Global Tectonics*, Wiley-Blackwell, Chichester, pp. 370–378.

14. Santosh, M., 2010, 'Supercontinent tectonics and biogeochemical cycle: A matter of "life and death"', *Geoscience Frontiers*, 1 (1), pp. 21–30.

15. Blankenship, R.E., 2010, 'Early evolution of photosynthesis', *Plant Physiology*, 154 (2), pp. 434–38.

16. Walker, J., 1978, 'The early history of oxygen and ozone in the atmosphere', *Pure and Applied Geophysics*, 117 (3), pp. 498–512.

17. Rich, M., et al., 2021, 'Lipid exchanges drove the evolution of mutualism during plant terrestrialization', *Science*, 372 (6544), pp. 864–8.

18. Harrison, C. J., and Morris, J.L., 2018, 'The origin and early evolution of vascular plant shoots and leaves', *Philosophical Transactions of the Royal Society of London. Series B, Biological Sciences*, 373 (1739), pp. 1-14.

19. Weng, J., and Chapple, C., 2010, 'The origin and evolution of lignin biosynthesis', *The New Phytologist*, 187 (2), pp. 273–85.

## Chapter 5

1. Seward, A.C., 2010, 'Preservation of plants as fossils', in *Plant Life Through the Ages: A geological and botanical retrospect*, Cambridge University Press, Cambridge, pp. 45–59.

2. Rickards, R.B., 2000, 'The age of the earliest club mosses: The Silurian Baragwanathia flora in Victoria, Australia', *Geological Magazine*, 37 (2), pp. 207–9.

3. Gensel, P.G., 2008, 'The earliest land plants', *Annual Review of Ecology, Evolution, and Systematics*, 39, pp. 459–77.

4. Rubinstein, C.V., et al., 2010, 'Early middle Ordovician evidence for land plants in Argentina (eastern Gondwana)', *The New Phytologist*, 188 (2), pp. 365–9.

5. Key, R., et al., 2015, 'The Zambezi River: An archive of tectonic events linked to the amalgamation and disruption of Gondwana and subsequent evolution of the African Plate', *South African Journal of Geology: Being the Transactions of the Geological Society of South Africa*, 118 (4), pp. 425–38.

6. Hueber, F.M., 2001, 'Rotted Wood-alga-fungus: The history and life of prototaxites Dawson 1859', *Review of Palaeobotany and Palynology*, 116 (1), pp. 123–58.

7. Torsvik, T.H., and Cocks, L.R.M., 2004, 'Earth geography from 400 to 250 MA: A palaeomagnetic, faunal and facies review', *Journal of the Geological Society, London*, 161 (4), pp. 555–72.

8. Beerling, D.J., Osborne, C.P., and Chaloner. W.G., 2001, 'Evolution of leaf-form in land plants linked to atmospheric $CO_2$ decline in the late Palaeozoic Era', *Nature*, 410 (6826), pp. 352–4.

9. Pawlik, L., et al., 2020, 'Impact of trees and forests on the Devonian landscape and weathering processes with implications to the global Earth's system properties: A critical review', *Earth-Science Reviews*, 205, p. 103200.

10. Sperry, J.S., 2003, 'Evolution of water transport and xylem structure', *International Journal of Plant Sciences*, 164 (S3), pp. S115–27.

11. Gerrienne, P., et al., 2011, 'A simple type of wood in two early Devonian plants', *Science*, 333 (6044), p. 837.

12. Gibling, M.R., et al., 2014, 'Palaeozoic co-evolution of rivers and vegetation: A synthesis of current knowledge', *Proceedings of the Geologists' Association*, 125 (5–6), pp. 524–33.

13. Virginia Tech, 1999, 'Earliest modern tree lived 360–345 million years ago', *Science Daily*, 22 April, https://www.sciencedaily.com/releases/1999/04/990422060147

14. www.sci.waikato.ac.nz/evolution/EvolutionOfLife.shtml

15. Algeo, T.J., 1998, 'Terrestrial-marine teleconnections in the Devonian: Links between the evolution of land plants, weathering processes, and marine anoxic events', *Philosophical Transactions of the Royal Society of London. Series B, Biological Sciences*, 353 (1365), pp. 113.

16. Sallan, L., and Coates, M., 2010, 'End-Devonian extinction and a bottleneck in the early evolution of modern jawed vertebrates', *Proceedings of the National Academy of Sciences of the United States of America*, 107 (22), pp. 10131–35.

17. Gouillard, J., 2004, 'Histoire des entomologistes français, 1750–1950', *Édition Entièrement Revue et Augmentée*', Boubée, Paris, p. 287.

18. Harrison, J., Kaiser, A., and VandenBrooks, J., 2010, 'Atmospheric oxygen level and the evolution of insect body size', *Proceedings. Biological Sciences/The Royal Society*, 277 (1690), pp. 1937–46.

19. Cleal, C.J., and Thomas, B., 2005, 'Palaeozoic tropical rainforests and their effect on global climates: Is the past the key to the present?' *Geobiology*, 3 (1), pp. 13–31.

20. Scott, A.C., and Taylor, T., 1983, 'Plant/animal interactions during upper Carboniferous', *The Botanical Review; Interpreting Botanical Progress*, 49 (3), pp. 259–307.

21. Glasspool, I.J., et al., 2015, 'The impact of fire on the late Paleozoic Earth system', *Frontiers in Plant Science*, 6 (September), p. 756.

22. Nelsen, M., et al., 2016, 'Delayed fungal evolution did not cause the Paleozoic peak in coal production', *Proceedings of the National Academy of Sciences of the United States of America*, 113 (9), pp. 2442–7.

23. Feulner, G., 2017, 'Formation of most of our coal brought Earth close to global glaciation', *Proceedings of the National Academy of Sciences of the United States of America*, 114 (43), pp. 11333–7.

24. Nance, R., et al., 2012, 'A brief history of the Rheic Ocean', *Geoscience Frontiers*, 3 (2), pp. 125–35.

25. Correia, P., and Murphy, J.B., 2020, 'Iberian–Appalachian connection is the missing link between Gondwana and Laurasia that confirms a Wegenerian Pangaea configuration', *Scientific Reports*, 10 (1), p. 2498.

26. Parrish, J., 1993., 'Climate of the supercontinent Pangaea', *Journal of Geology*, 101 (2), pp. 215–33.

27. Sahney, S., Benton, M., and Falcon-Lang, H., 2010, 'Rainforest collapse triggered Carboniferous tetrapod diversification in Euramerica', *Geology*, 38 (12), pp. 1079–82.

28. Yashina, S., et al., 2012, 'Regeneration of whole fertile plants from 30,000-year-old fruit tissue buried in Siberian permafrost', *Proceedings of the National Academy of Sciences of the United States of America*, 109 (10), pp. 4008–13.

29. Pires, N.D., 2014, 'Seed evolution: Parental conflicts in a multi-generational household', *Biomolecular Concepts*, 5 (1), pp. 71–86.

30. Perrin, R. et al., 2005, 'Gravity signal transduction in primary roots', *Annals of Botany*, 96 (5), pp. 737–43.

31. Weijers, D., Jennifer N., and Zhenbiao Y., 2018, 'Auxin: Small molecule, big impact', *Journal of Experimental Botany*, 69 (2), pp. 133–6.

32. Sheldrake, M., 2020, *Entangled Life: How fungi make our worlds, change our minds and shape our futures*, The Bodley Head, London, pp. 165–93.

33. Kirkham, M.B., 2014, 'Stomatal anatomy and stomatal resistance', *Principles of Soil and Plant Water Relations*, 2nd edition, edited by M.B. Kirkham, Academic Press, Boston, pp. 431–51.

34. Simard, S., and Durall, D., 2004, 'Mycorrhizal networks: A review of their extent, function, and importance', *Canadian Journal of Botany*, 82 (8), pp. 1140–65.

## Chapter 6

1. Bowen, E.G., 1966, 'The Welsh colony in Patagonia 1865–1885: A study in historical geography', *Geographical Journal*, 132 (1), pp. 16–27.

2. Pant, D.D., 1987, 'Reproductive biology of the Glossopterids and their affinities', *Bulletin de la Société Botanique de France*, 134 (2), pp. 77–93.

3. Cariglino, B., 2015, 'New Glossopterid polysperms from the Permian La Golondrina Formation (Santa Cruz Province, Argentina): Potential affinities and biogtratigraphic implications', *Sociedade Brasileira de Paleontologia*, pp. 379–90.

4. Barghusen, H.R., 1975, 'A review of fighting adaptations in Dinocephalians (Reptilia, Therapsida)', *Paleobiology*, 1 (3), pp. 295–311.

5. Benoit, J., et al., 2021, 'Palaeoneurology and palaeobiology of the Dinocephalian Therapsid Anteosaurus Magnificus', *Acta Palaeontologica Polonica*, 66 (1), pp. 2021.

6. Edirisooriya, G., Dharmagunawardhane, H.A., and Mcloughlin, S., 2018, 'The first record of the Permian *Glossopteris* flora from Sri Lanka: Implications for hydrocarbon source rocks in the Mannar Basin', *Geological Magazine*, 155 (4), pp. 907–20.

7. Pigg, K.B., and Taylor, T., 1993, 'Anatomically preserved *Glossopteris* stems with attached leaves from the Central Transantarctic Mountains, Antarctica', *American Journal of Botany*, 80 (5), pp. 500–16.

8. Neish, P.G., Drinnan, A., and Cantrill, D., 1993, 'Structure and ontogeny of *Vertebraria* from silicified Permian sediments in East Antarctica', *Review of Palaeobotany and Palynology*, 79 (3), pp. 221–43.

9. McLoughlin, S., 2012, '*Glossopteris*: Insights into the architecture and relationships of an iconic Permian Gondwanan plant', *J. Botan. Soc. Bengal*, 65, pp. 1–14.

10. Gulbranson, E.L., et al., 2014, 'Leaf habit of late Permian *Glossopteris* trees from high-palaeolatitude forests', *Journal of the Geological Society*, 171 (4), pp. 493–507.

11. Decombeix, A., Taylor, E.L., and Taylor, T., 2016, 'Bark anatomy of late Permian glossopterid trees from Antarctica', *IAWA Journal/International Association of Wood Anatomists*, 37 (3), pp. 444–58.

12. Decombeix, A., Taylor, E.L., and Taylor, T., 2010, 'Epicormic shoots in a Permian gymnosperm from Antarctica', *International Journal of Plant Sciences*, 171 (7), pp. 772–82.

13. Murthy, S., et al., 2020, 'Evidence of recurrent wildfire from the Permian coal deposits of India: Petrographic, scanning electron microscopic and palynological analyses of fossil charcoal', *Palaeoworld*, 29 (4), pp. 715–28.

14. White, M., 1986, *The Greening of Gondwana: The 400 million year story of Australia's plants*, Reed Books, Sydney, pp. 108–119.

15. Nishida, H., et al., 2004, 'Zooidogamy in the late Permian genus *Glossopteris*', *Journal of Plant Research*, 117 (4), pp. 323–28.

16. Bernardi, M., et al., 2017, 'Late Permian (Lopingian) terrestrial ecosystems: A global comparison with new data from the low-latitude Bletterbach biota', *Earth-Science Reviews*, 175 (December), pp. 18–43.

17. Cúneo, N., et al., 1993, 'The *Glossopteris* flora in Antarctica: Taphonomy and paleoecology', *Comptes rendus de l'Académie des Sciences, 2 (January), pp. 13–40.*

18. Slater, B., McLoughlin, S., and Hilton, J., 2012, 'Animal-plant interactions in a middle Permian permineralised peat of the Bainmedart coal measures, Prince Charles Mountains, Antarctica', *Palaeogeography, Palaeoclimatology, Palaeoecology*, 363–364 (November), pp. 109–26.

19. Parrish, J., 1993, 'Climate of the Supercontinent Pangaea', *Journal of Geology*, 101 (2), pp. 215–33.

20. Arraut, J., et al., 2012, 'Looking at large-scale moisture transport and its relation to Amazonia and to subtropical rainfall in South America', *Journal of Climate*, 25 (2), pp. 543–56.

21. Day, M., et al., 2015, 'When and how did the terrestrial mid-Permian mass extinction occur? Evidence from the tetrapod record of the Karoo

Basin, South Africa', *Proceedings. Biological Sciences/The Royal Society*, 282 (1811), pp. 1–9.

22. Zavada, M.S., and Mentis, M.T., 1992, 'Plant-animal interaction: The effect of Permian megaherbivores on the glossopterid flora', *American Midland Naturalist*, 127 (1), pp. 1–12.

## Chapter 7

1. Freedman, L., 1982, 'The war of the Falkland Islands, 1982', *Foreign Affairs*, 61 (1), pp. 196–210.

2. Du Toit, A.L., 1937, *Our Wandering Continents*, Oliver and Boyd, Edinburgh, pp. 118–120.

3. Adie, Raymond J., 1952, 'The position of the Falkland Islands in a reconstruction of Gondwanaland', *Geological Magazine*, 89 (6), pp. 401–10.

4. Royle, Stephen A., 1985, 'The Falkland Islands, 1833–1876: The establishment of a colony', *The Geographical Journal*, 151 (2), pp. 204–14.

5. Aldiss, D.T., and Edwards, E.J., 1999, 'The geology of the Falkland Islands', *British Geological Survey Technical Report*, WC/99110, p. 3.

6. Seward, A.C., 1912, 'The *Glossopteris* flora', *The New Phytologist*, 11 (1), pp. 33–6.

7. Scotese, C. R., and Langford, R.P., 1995, 'Pangaea and the paleogeography of the Permian', *The Permian of Northern Pangaea: Volume 1: Paleogeography, paleoclimates, stratigraphy*, Springer, Berlin/Heidelberg, pp. 3–19.

8. Benton, M., 2005, *When Life Nearly Died: The greatest mass extinction of all time*, Thames and Hudson, London, pp. 180–203.

9. Rayner, R.J., 1992, 'Phyllotheca: The pastures of the late Permian', *Palaeogeography, Palaeoclimatology, Palaeoecology*, 92 (1), pp. 31–40.

10. Bernardi, M., et al., 2017.

11. Morón, S., et al., 2019, 'Long-lived transcontinental sediment transport pathways of East Gondwana', *Geology*, 47 (6), pp. 513–16.

12. Ghosh, S.C., 2002, 'The Raniganj Coal Basin: An example of an Indian Gondwana rift', *Sedimentary Geology*, 147 (1), pp. 155–76.

13. Slater, B., McLoughlin, S., and Hilton, J., 2015, 'A high-latitude Gondwanan lagerstätte: The Permian permineralised peat biota of the Prince Charles Mountains, Antarctica', *Gondwana Research*, 27 (4), pp. 1446–73.

14. Viglietti, P., et al., 2016, 'The *Daptocephalus* Assemblage Zone (Lopingian), South Africa: A proposed biostratigraphy based on a new compilation of stratigraphic ranges', *Journal of African Earth Sciences*, 113 (January), pp. 153–64.

15. Robertson, A., Campbell, H.J., Johnston, M., and Mortimer, N., 2019, 'Introduction to Paleozoic–Mesozoic geology of South Island, New Zealand: Subduction-related processes adjacent to SE Gondwana', *Geological Society, London, Memoirs*, 49 (1), pp. 1–14.

16. Isozaki, Y., 2009, 'Illawarra reversal: The fingerprint of a superplume that triggered Pangaean break-up and the end-Guadalupian (Permian) mass extinction', *Gondwana Research*, 15 (3–4), pp. 421–32.

17. Urban, J., et al., 2017, 'Increase in leaf temperature opens stomata and decouples net photosynthesis from stomatal conductance', *Pinus Taeda* and *Populus Deltoides X Nigra. Journal of Experimental Botany*, 68 (7), pp. 1757–67.

## Chapter 8

1. Nowak, H., Vérard, C., and Kustatscher, E., 2020, 'Palaeophytogeographical patterns across the Permian–Triassic boundary', *Frontiers of Earth Science in China*, 8, p. 609.

2. Alvarez, W., 1997, *T-Rex and the Crater of Doom*, Penguin, London, pp. 59–81.

3. Chiarenza, A., et al., 2020, 'Asteroid impact, not volcanism, caused the end-Cretaceous dinosaur extinction', *Proceedings of the National Academy of Sciences of the United States of America*, 117 (29), pp. 17084–93.

4. Benton, M., 2005, pp. 1–366.

5. Botha, J., 2020, 'The paleobiology and paleoecology of South African *Lystrosaurus*', *PeerJ*, 8 (November), p. 10408.

6. Ward, P.D., Montgomery, D.R., and Smith. R., 2000, 'Altered river morphology in South Africa related to the Permian-Triassic extinction', *Science*, 289 (5485), pp. 1740–43.

7. Ivanov, A., et al., 2013, 'Siberian Traps large igneous province: Evidence for two flood basalt pulses around the Permo-Triassic Boundary and in the middle Triassic, and contemporaneous granitic magmatism', *Earth-Science Reviews*, 122 (July), pp. 58–76.

8. Augland, L.E., et al., 2019, 'The main pulse of the Siberian Traps expanded in size and composition', *Scientific Reports*, 9 (1), p. 18723.

9. Ivanov, A., 2007, 'Evaluation of different models for the origin of the Siberian Traps', *Special Papers-Geological Society Of America*, pp. 430, 669.

10. Benton, M.J., 2018, 'Hyperthermal-driven mass extinctions: Killing models during the Permian-Triassic mass extinction', *Philosophical Transactions. Series A, Mathematical, Physical, and Engineering Sciences*, 376, p. 2130.

11. Retallack, G., and Jahren, A.H., 2008, 'Methane release from igneous intrusion of coal during late Permian extinction events', *Journal of Geology*, 116 (1), pp. 1–20.

12. Brand, U., et al., 2016, 'Methane hydrate: Killer cause of Earth's greatest mass extinction', *Palaeoworld*, 25 (4), pp. 496–507.

13. Ward, P., 2008, *Under a Green Sky: Global warming, the mass extinctions of the past, and what they can tell us about our future*, Harper Collins, New York, pp. 1–256.

14. Song, H., et al., 2021, 'Thresholds of temperature change for mass extinctions', *Nature Communications*, 12 (1), p. 4694.

15. Clarkson, M.O., et al., 2015, 'Ocean acidification and the Permo-Triassic mass extinction', *Science*, 348 (6231), pp. 229–32.

16. Benca, J., et al., 2018, 'UV-B-induced forest sterility: Implications of ozone shield failure in Earth's largest extinction', *Science Advances*, 4 (2), p. e1700618.

17. Visscher, H., et al., 2004, 'Environmental mutagenesis during the end-Permian ecological crisis', *Proceedings of the National Academy of Sciences of the United States of America*, 101 (35), pp. 12952–6.

18. Glasspool, I., et al., 2015, 'The impact of fire on the late Paleozoic Earth system', *Frontiers in Plant Science*, 6 (September), p. 756.

19. Retallack, G., Veevers, J., and Morante, R., 1996, 'Global coal gap between Permian-Triassic extinction and middle Triassic recovery of peat-forming plants', *GSA Bulletin*, 108 (2), p. 195–207.

20. Shen, Shu-Zhong, et al., 2019, 'A sudden end-Permian mass extinction in South China', *GSA Bulletin*, 131 (1–2), pp. 205–23.

21. Fielding, C.R., Frank, T., McLoughlin, S., et al., 2019, 'Age and pattern of the southern high-latitude continental end-Permian extinction constrained by multiproxy analysis', *Nature Communications*, 10 (1), p. 385.

22. Payne, J., and Krump, L., 2007, 'Evidence for recurrent early Triassic massive volcanism from quantitative interpretation of carbon isotope fluctuations', *Earth and Planetary Science Letters*, 256 (1), pp. 264–77.

23. Retallack, G.J., 1995, 'Permian-Triassic life crisis on land', *Science*, 267 (5194), pp. 77–80.

24. Benton, M., Forth, J., and Langer, M., 2014, 'Models for the rise of the dinosaurs', *Current Biology: CB*, 24 (2), pp. r87–95.

25. Benton, M., 2016, 'The Triassic', *Current Biology: CB*, 26 (23), pp. r1214–18.

26. Chatterjee, S.K., and Scotese, C.R, 1999, 'The break-up of Gondwana and the evolution and biogeography of the Indian Plate', *Proceedings of the Indian Academy of Sciences: Section A Part 3 Mathematical Sciences*, 65A (3), pp. 397–425.

27. Atkinson, B., et al., 2018, 'Additional evidence for the Mesozoic diversification of conifers: Pollen cone of *Chimaerostrobus Minutus* Gen. et Sp. Nov. (Coniferales), from the lower Jurassic of Antarctica', *Review of Palaeobotany and Palynology*, 257 (October), pp. 77–84.

28. Stockey, Ruth A., 1982, 'The Araucariaceae: An evolutionary perspective', *Review of Palaeobotany and Palynology*, 37 (1), pp. 133–54.

29. Hammer, W.R., and Hickerson, W.J., 1994, 'A crested theropod dinosaur from Antarctica', *Science*, 264 (5160), pp. 828–30.

30. Bell, Phil R., Snively, E., 2000, 'Polar dinosaurs on parade: A review of dinosaur migration', *Alcheringa: An Australasian Journal of Palaeontology*, 32 (3), pp. 271–84.

31. Condamine, F., et al., 2020, 'The rise of angiosperms pushed conifers to decline during global cooling', *Proceedings of the National Academy of Sciences of the United States of America*, 117 (46), pp. 28867–75.

32. Wilson Mantilla, G., et al., 2021, 'Earliest Palaeocene purgatoriids and the initial radiation of stem primates', *Royal Society Open Science*, 8 (2), p. 210050.

33. Thornhill, A., et al., 2015, 'Interpreting the modern distribution of *Myrtaceae* using a dated molecular phylogeny', *Molecular Phylogenetics and Evolution*, 93 (December), pp. 29–43.

34. Johnson, L.A.S., and Briggs, B., 1975, 'On the *Proteaceae* — the evolution and classification of a southern family', *Botanical Journal of the Linnean Society. Linnean Society of London*, 70 (2), pp. 83–182.

35. Feild, T., Zwieniecki, M., and Holbrook, N., 2000, '*Winteraceae* evolution: An ecophysiological perspective', *Annals of the Missouri Botanical Garden. Missouri Botanical Garden*, 87 (3), pp. 323–34.

36. Rowley, D., 1996, 'Age of initiation of collision between India and Asia: A review of stratigraphic data', *Earth and Planetary Science Letters*, 145 (1), pp. 1–13.

37. Tada, R., Zheng, H., and Clift, P., 2016, 'Evolution and variability of the Asian monsoon and its potential linkage with uplift of the Himalaya and Tibetan plateau', *Progress in Earth and Planetary Science*, 3 (1), pp.1–26.

38. Burnham, R., and Johnson, K., 2004, 'South American palaeobotany and the origins of neotropical rainforests', *Philosophical Transactions of the Royal Society of London. Series B, Biological Sciences*, 359 (1450), pp. 1595–610.

39. Toumoulin, A., et al., 2020, 'Quantifying the effect of the Drake Passage opening on the Eocene Ocean', *Paleoceanography and Paleoclimatology*, 35 (8), pp. 1–37.

40. Strullu-Derrien, C., Selosse, M., Kenrick, P., and Martin., F., 2018, 'The origin and evolution of mycorrhizal symbioses: From palaeomycology to phylogenomics', *New Phytologist*, 220 (4), pp. 1012–30.

41. Cook, Lyn G., Crisp., M., 2005, 'Not so ancient: The extant crown group of *Nothofagus* represents a post-Gondwanan radiation', *Proceedings. Biological Sciences/The Royal Society*, 272 (1580), pp. 2535–44.

42. Biffin, E., Conran, J., and Lowe, A., 2011, 'Podocarp evolution: A molecular phylogenetic perspective', *Ecology of the Podocarpaceae in Tropical Forests*, 95, pp. 1–21.

43. Strömberg, C.A., 2011, 'Evolution of grasses and grassland ecosystems', *Annual Review of Earth and Planetary Sciences*, 39 (1), pp. 517–44.

44. Bobe, R., and Behrensmeyer, A., 2004, 'The expansion of grassland ecosystems in Africa in relation to mammalian evolution and the origin of the genus Homo', *Palaeogeography, Palaeoclimatology, Palaeoecology*, 207 (3), pp. 399–420.

45. Mortimer, N., and Campbell, H.J., 2014, *Zealandia: Our continent revealed*, Penguin, Auckland, pp. 158–167.

## Chapter 9

1. Mortimer, N., and Campbell, H.J., 2014.

2. Turnbull, I., and Allibone, A., 2003, 'Geology of the Murihiku area', *Institute of Geological and Nuclear Sciences NZ*, 20, pp. 15–25.

3. Landis, C.A., Campbell, H.J., et al., 1999, 'Permian-Jurassic strata at Productus Creek, Southland, New Zealand: Implications for terrane dynamics of the eastern Gondwanaland margin', *New Zealand Journal of Geology and Geophysics*, 42 (2), pp. 255–78.

4. Christenhusz, M., and Byng, J., 2016, 'The number of known plants species in the world and its annual increase', *Phytotaxa*, 261 (3), pp. 201–17.

5. Ehrenberg, R., 2015, 'Global forest survey finds trillions of trees', *Nature*, (September), 525, pp. 170–71.

6. Beech, E., Rivers, M., Oldfield, S., and Smith, P., 2017, 'GlobalTreeSearch: The first complete global database of tree species and country distributions', *Journal of Sustainable Forestry*, 36 (5), pp. 454–89.

7. Smillie, M.J., 2016, 'The New Zealand Tree Register', https://register.notabletrees.org.nz/tree/view/1375

8. Domec, J., et al., 2008, 'Maximum height in a conifer is associated with conflicting requirements for xylem design', *Proceedings of the National Academy of Sciences of the United States of America*, 105 (33), pp. 12069–74.

9. Cardoso, D., et al., 2017, 'Amazon plant diversity revealed by a taxonomically verified species list', *Proceedings of the National Academy of Sciences of the United States of America*, 114 (40), pp. 10695–700.

10. DeAngelis, D.L., 2008, 'Boreal forest', *Encyclopedia of Ecology*, edited by Sven Erik Jørgensen and Brian D. Fath, Academic Press, Oxford, pp. 493–5.

11. Gibbs, G., 2006, *Ghosts of Gondwana*, Craig Potton Publishing, Nelson, pp. 138–44.

12. Bunce, M., et al., 2005. 'Ancient DNA provides new insights into the evolutionary history of New Zealand's extinct giant eagle', *PLoS Biology*, 3 (1) p. e9.

13. Knapp, M., et al., 2005, 'Relaxed molecular clock provides evidence for long-distance dispersal of *Nothofagus* (southern beech)', *PLoS Biology*, 3 (1), p. e14.

14. Barker, N., et al., 2007, 'Molecular dating of the "Gondwanan" plant family *Proteaceae* is only partially congruent with the timing of the break-up of Gondwana', *Journal of Biogeography*, 34 (12), pp. 2012–27.

15. Liggins, L., et al., 2008, 'Origin and post-colonization evolution of the Chatham Islands skink (oligosoma nigriplantare nigriplantare)', *Molecular Ecology*, 17 (14), pp. 3290–305.

16. Baker, Allan J., et al., 2014, 'Genomic support for a moa-tinamou clade and adaptive morphological convergence in flightless ratites', *Molecular Biology and Evolution*, 31 (7), pp. 1686–96.

17. Jin, S., and Wang, J., 2008, 'Spreading change of Africa-South America plate: Insights from space geodetic observations', *International Journal of Earth Sciences*, 97 (6), pp. 1293–300.

## Chapter 10

1. Goswami, S., 2015, 'Impact of coal mining on environment: A study of Raniganj and Jharia coal field in India', *IAFOR Journal of Arts & Humanities*, 3 (1), pp. 2–17.

2. Bera, R., 2020, 'For a fistful of coal', *The Week*, 27 November, https://www.theweek.in/theweek/cover/2020/11/26/for-a-fistful-of-coal.html

3. Singh, G., 2019, 'The burning coalfields of Jharia belch poison for local residents', Mongabay, 3 October, https://india.mongabay.com/2019/10/the-burning-coalfields-of-jharia-belch-poison-for-local-residents/

4. Singh, K.M., et al. 2012, 'Rural poverty in Jharkhand, India: An empirical study based on panel data', ICAR Research Complex for eastern region, Patna, https://mpra.ub.uni-muenchen.de/45258/1/MPRA_paper_45258.pdf

5. Dey, S., 2021, 'Jharia most polluted place in India in 2018, Greenpeace report', *Hindustan Times*, 21 January.

6. Mishra, H.K., Chandra, T.K., and Verma, R.P., 1990, 'Petrology of some Permian coals of India', *International Journal of Coal Geology*, 16 (1), pp. 47–71.

7. Chakravarty, S., and Somanathan, E., 2021, 'There is no economic case for new coal plants in India', *World Development Perspectives*, 24 (December), p. 100373.

8. Lin, F., Inglesi-Lotz, R., and Chang, T., 2018, 'Revisit coal consumption, $CO_2$ emissions and economic growth nexus in China and India using a newly developed bootstrap ARDL bound test', *Energy Exploration & Exploitation*, 36 (3), pp. 450–63.

9. Clarke, J., 2001, 'Indian-Australian coal ties tighten', *Argus Media*, 27 September.

10. Mohr, S., et al., 2011, 'Projection of Australian coal production: Comparisons of four models, *International Journal of Coal Geology*, 86(4), pp. 329–341.

11. Australian government website: https://www.ga.gov.au/scientific-topics/energy/province-sedimentary-basin-geology/petroleum/onshore-australia/bowen-basin. Accessed 30 November 2021.

12. NSW government website: https://www.business.nsw.gov.au/industry-sectors/industry-opportunities/mining-and-resources/coal/coal-in-nsw. Accessed 30 November 2021.

13. Jones, D., 2021, 'EMBER global electricity review', p.17.

14. Holland, M., 2017, 'Health impacts of coal fired power plants in South Africa', *Coal Kills*, pp.4–5.

15. Akinbami, O., Oke, S., and Bodunrin, M., 2021, 'The state of renewable energy development in South Africa: An overview', *Alexandria Engineering Journal*, 60 (6), pp. 5077–93.

16. Kholod, N., et al., 2020, 'Global methane emissions from coal mining to continue growing even with declining coal production', *Journal of Cleaner Production*, 256, p. 120489.

17. Jakob, M., et al., 2020, 'The future of coal in a carbon-constrained climate', *Nature Climate Change*, 10 (8), pp. 704–7.

18. US Department of Commerce, NOAA, and Global Monitoring Laboratory, 2005, 'Global Monitoring Laboratory: Carbon Cycle Greenhouse Gases'.

19. Delmotte, V., 2021, 'Climate change 2021: Summary for policymakers', *Intergovernmental Panel on Climate Change*, p. 6.

20. Li, Yao-Jun, Yong-Jian Ding, Dong-Hui Shangguan, and Rong-Jun Wang, 2019, 'Regional differences in global glacier retreat from 1980 to 2015', *Advances in Climate Change Research*, 10 (4), pp. 203–13.

21. Wunderling, N., 2020, 'Global warming due to loss of large ice masses and Arctic summer sea ice', *Nature Communications*, 11 (1), p. 5177.

22. Wu, Y., et al., 2021, 'Sixfold increase of atmospheric $pCO_2$ during the Permian-Triassic mass extinction', *Nature Communications*, 12 (1), p. 2137.

23. Song, H., 2021, 'Thresholds of temperature change for mass extinctions', *Nature Communications*, 12 (1), p. 4694.

24. Weisse, M., and Goldman, L., 2021, 'Global Forest Watch Annual Report 2020'.

25. Williams, A., 2019, 'Observed impacts of anthropogenic climate change on wildfire in California', *Earth's Future*, 7 (8), pp. 892–910.

26. Biskaborn, B., et al., 2019, 'Permafrost is warming at a global scale', *Nature Communications*, 10 (1), p. 264.

27. Previdi, M., Smith, K., and Polvani, L., 2021, 'Arctic amplification of climate change: A review of underlying mechanisms', *Environmental Research Letters*, 16 (9), p. 093003.

28. Ceballos, G., Ehrlich, P., and Raven, P., 2020, 'Vertebrates on the brink as indicators of biological annihilation and the sixth mass extinction', *Proceedings of the National Academy of Sciences of the United States of America*, 117 (24), pp. 13596–602.

29. Gardner, A., et al., 2021, 'Is photosynthetic enhancement sustained through three years of elevated $CO_2$ exposure in 175-year-old Quercus robur?' *Tree Physiology*, 42 (1), pp. 130-144.

30. Song, X., et al., 2018, 'Global land change from 1982 to 2016', *Nature*, 560 (7720), pp. 639–43.

31. Fei, S., et al., 2017, 'Divergence of Species Responses to climate change', *Science Advances*, 3 (5), p. e1603055.

32. Hansson, A., Dargusch, P., and Shulmeister, J., 2021, 'A review of modern treeline migration, the factors controlling it and the implications for carbon storage', *Journal of Mountain Science*, 18 (2), pp. 291–306.

33. Lovelock, J., et al., 1979, *Gaia: A new look at life on Earth*, Oxford University Press, Oxford, pp. 1–141.

34. Retallack, G., 2021, 'Soil carbon dioxide planetary thermostat', *Astrobiology*, 21 (10), pp. 1–8.

35. Swann, A., et al., 2010, 'Changes in Arctic vegetation amplify high-latitude warming through the greenhouse effect', *Proceedings of the National Academy of Sciences of the United States of America*, 107 (4), pp. 1295–1300.

# ILLUSTRATION CREDITS

**pages viii–1:** Alinabel/Shutterstock

**page 5:** Auckland Libraries Heritage Collections 5A-18, photographer Henry Bowers

**page 10:** Auckland Libraries Heritage Collections 5A-34, photographer Henry Bowers

**page 16:** Ewan Fordyce

**page 20:** Pictorial Press Ltd/Alamy Stock Photo

**page 21:** Chuck Place/Alamy Stock Photo

**page 25:** William Smith/British Library

**page 28:** Alamy Stock Photo

**page 31:** Georges Cuvier, Alexandre Brongniart, Essai minéralique sur les environs de Paris (1st edition 1808)

**pages 38–39:** Paulina Barry

**page 50:** Science Photo Library/Alamy Stock Photo

**page 54:** Photo Vault/Alamy Stock Photo

**page 55:** Sam van Der Weerden

**page 57:** Manish Lakhani/Alamy

**page 61:** Royal Botanic Gardens, Kew

**page 64:** Science Photo Library/Alamy Stock Photo

**page 70:** Stockholm Maritime Museum

**page 74:** Bill Morris

**page 76:** Alfred Wegener Institute for Polar and Marine Research

**page 77:** United States Geological Survey

**page 83:** Paulina Barry

**page 95:** United States Geological Survey

**page 96:** K. D. Schroeder, Subduction-en.svg from Wikimedia Commons, CC-BY-SA 4.

**page 101:** Yvonne Baur/Shutterstock

**page 104:** Paulina Barry

**page 108:** Cameron Brideoake

**page 113:** Paulina Barry

**page 115:** Bodor Tivadar/Shutterstock

**pages 116–117:** Paulina Barry

**page 119:** Sean Pavone/Alamy Stock Photo

**pages 126–127:** Paulina Barry

**page 133:** Bill Morris

**page 141:** Paulina Barry

**page 147:** Bill Morris

**pages 148–149:** Paulina Barry

**page 160:** Bill Morris

**page 165:** Paulina Barry

**pages 166–167:** Paulina Barry

**page 179:** Ryz Haydul/Alamy Stock Photo

**page 193:** Bill Morris

**page 197:** Ashley Cooper/Alamy Stock Photo

**page 212:** Joerg Boethling/Alamy Stock Photo

# INDEX